EDUCATION FOR MATHEMATICS IN THE WORKPLACE

Mathematics Education Library

VOLUME 24

The titles published in this series are listed at the end of this volume.

EDUCATION FOR MATHEMATICS IN THE WORKPLACE

Edited by

ANNIE BESSOT
Laboratoire Leibniz,
Université Joseph Fourier,
Grenoble, France

and

JIM RIDGWAY
School of Education,
University of Durham,
United Kingdom

SPRINGER-SCIENCE+BUSINESS MEDIA, B.V.

Library of Congress Cataloging-in-Publication Data

ISBN 978-90-481-5600-9 ISBN 978-0-306-47226-8 (eBook)
DOI 10.1007/978-0-306-47226-8

Printed on acid-free paper

TABLE OF CONTENTS

ACKNOWLEDGMENT

We would like to thank Rosemary Rodwell for help translating several chapters into English.

Annie Bessot
Jim Ridgway

INTRODUCTION

GÉRARD VERGNAUD

The idea of this book was born in the discussion group 'Mathematics in the workplace' at the International Congress on Mathematics education, in Sevilla (1996). This idea has been developed at different occasions since that time. The book is made of different kinds of contributions:

-ethnograhic or quasi-ethnographic studies on the use of mathematics at the workplace (Zevenbergen; Noss, Hoyles and Pozzi);

-analytical descriptions of the mathematical activities involved in different professions (Bessot; Eberhard; Mercier);

-presentations and explanations concerning the kind of mathematics that is taught in vocational schools (Straesser; Wedege; Hahn), or the kind of mathematics that is needed and should be taught better (Ridgway ; Wake and Williams; Gillespie).

Finally some chapters deal with specific didactical problems like slow learners (Van der Zwaart), technological resources, or symbolic tools (Evans). Several contributions deal with the problem of articulating mathematics taught in schools and mathematics used in professional practices. Evans proposes to reformulate the problem of transfer by relating with one another different practices and different kinds of signifiers. It is also a good point.

The most important idea that comes out of the studies presented in this book is that mathematics is above all an activity. This activity takes place in all kinds of situations and workplaces, as different as farming, building, nursing, driving a plane, or designing a satellite.

A strong characteristic is that mathematics is all embedded in these activities, and not easily made explicit by the professionals who are asked to explain how they do what they have to do. Most of the knowledge used at the workplace remains implicit, sometimes unconscious. This is true even for high-level educated professionals like pilots, medical doctors or engineers. It is even true for researchers in mathematics themselves, in the sense that they write papers that do not convey all the mathematical ideas they can think of, while they elaborate new views and proofs. Therefore we need a theoretical framework to grasp the relation between activity and conceptualisation. I will present a theory on this problem further on, with the concept of scheme.

Bessot & Ridgway (eds.), Education for Mathematics in the Workplace, xvii-xxiv.

Of course the implicit character of a major part of our cognitive processes does not concern mathematics only, but all domains of activity, including the production and processing of language itself. Five-year old children produce syntactically correct sentences in their mother language without being aware of the rules they use.

Another important idea is that several mathematical domains appear again and again in the routines used at the workplace: proportionality, graph reading, map-reading, evaluation and approximation… This point is clear in several chapters of the book; I will not develop here what is explained later with greater precision. But the fact that some parts of mathematics are used more often than others in many professions raises the question of the room given to them in the curriculum. Van der Zwaart, Wake and Williams address this problem most clearly, in dealing with slow learners and post-compulsary education. But it is a general question for mathematics education.

Routines must be analysed, as they stand at the core of a profession. But it is also fruitful to pay attention to non-routine activities: in industry and commerce to-day, expertise consists not only of being able to deal with familiar situations, but also to face new situations, never met before, and solve new problems. These non-routine activities raise several challenging questions for future research. What is expertise? And how is it acquired? How long does it take to become an expert? How is collective competence developed? How are mathematical ideas communicated inside a group? We know very little about these questions, except that it takes many years to become an expert, and that the competence of a group is more than the sum of the individual competencies inside the group.

An interesting point concerns errors. How are they generated? They are not usually taught; therefore they are invented. Invention does not necessarily mean sophisticated creativity, but also weakly controlled inference. A simple example is the 'additive percentage syndrome' studied by Hahn later in this book:

> If you add a tax of 18.6% to the price, in order to calculate the amount to be paid, then you subtract 18.6% from the amount paid to calculate the price

$$-x\% = \textit{inverse of } +x\%$$

This is typically a theorem-in-action, in spite of the fact that it is wrong; it is held to be true by many people.

It is sometimes easy to help people get rid of such errors, but there are also cases when they go on using the same wrong procedure, because the procedure is socially accepted: for instance sugar-cane workers in north-east Brazil go on calculating the area of quadrilaterals by multiplying the average length of two opposite sides by the average length of the other two opposite sides, even when the shape is far from being a rectangle. They also calculate the area of a triangle, with the same logic, as if it were a quadrilateral having lost one side (Acioly, 1994, 1995).

$$1/2(a + b) \times 1/2(c + 0)$$

Last but not least, professional competence, is the result of several different and interactive processes: initial education, experience, and in-service training. Experience is essential for the development of competencies. Its role must never be minimised. But it would also be a great mistake to consider as negligible the role of initial education and in service training: the technical and scientific sophistication of activity required at the workplace makes it more and more necessary not only to raise the level of initial education, and provide in-service training to employees, workers, and managers, at all levels of qualification and responsibility, but also to examine the relation between the mathematics taught in school and the mathematics used at work.

Education and training bring research in didactics and pedagogy to the scene. It is a wise and reasonable position that most authors in this book do not insist as much as some other authors do, on the gap and discrepancy between the knowledge taught in schools and universities, and the knowledge used in the workplace. There is no contradiction between them, and there are even many common points between what students find more natural to do in schools and what adults do at the workplace. Nevertheless, if one wants to make general education and professional training more operational, one is inevitably faced with the question of trying to transpose certain paradigmatic work situations into the class-room; also with the challenge of finding some equilibrium between the two functions of general education: transmission of mathematics as a scientific culture, and formation of a range of competencies likely to be useful in a variety of professions. Neither a purely utilitarian, nor a purely scientific conception of mathematics education are adequate.

Whatever essential the ethnographic approach may be (Zevenbergen), the challenge is not only to look at the workplace as it is, but also to create conditions for it to change. Things move rapidly to-day, in most branches of activity. Managers tend to think of change in terms of organisation of work, because organisation is the most visible factor of efficiency for them; but the time has come to study carefully the conditions under which people do what they have to do, and learn what they need to learn to keep on line with technological and social change. We learn a lot from experience, but experience does not consist only of great familiarity with a limited range of situations; it consists also of having met a variety of cases, and developed solutions for them (usually with the help of other persons, inside or outside the workplace).

Theoretical considerations

If mathematics at the workplace is an activity, we need an analytical theory of activity. Such expressions like 'situated cognition', 'community of practice', or 'activity' itself are not self-sufficient concepts. We need more.

Because we learn in situations, the concept of situation is essential: this is true at the workplace, also inside school. The theoretical reason for this is that knowledge is adaptation; it is to situations that we adapt ourselves. Consequently we need to identify the different classes of situations we have to deal with. It is too general to say that we adapt to reality, or to our environment, or even to objects. The same objects may have different properties in different situations: we may be easy with them in some cases and uneasy in other cases.

Because we need to grasp the process of adaptation itself, and because we adapt ourselves through our own activity, the concept of scheme is also essential: it specifies the concept of activity by delineating forms of organisation of activity related to specific classes of situations. Schemes are such forms.

Let me take two examples:

-The pilot described by Noss, Hoyles and Pozzi in their chapter, uses a 'nose' estimation of the angle of the wind with the runway and the air-plane trajectory, then a quadrant; instead of calculating some trigonometric function. He does not want to bother with such a calculation, even though he is perfectly able to use trigonometry.

-Bessot describes three different techniques used to construct a filler (in building industry); one of them is a pragmatic one, not algorithmic, still adequate most of the time.

Definition: *A scheme is an invariant organisation of activity and behaviour for a certain class of situations.*

This definition requires several comments:

1-It is the organisation that is invariant, not the behaviour. The same scheme can generate different behaviours for different situations belonging to the same class.

2-As a scheme addresses a class of situations, it is a universal, in the sense that one needs universal quantifiers to write the rules; yet schemes are usually local, and used for a limited range of values of the situation variables (domain, circumstances, conditions, numerical values etc.). This local and limited availability of schemes is a strong characteristic of mathematical activities at the workplace, as explained in several chapters.

3-Schemes are not usually algorithms; but algorithms form a sub-class of schemes.

Mathematicians have developed algorithms for many classes of situations; they concern geometry, arithmetical calculations, algebra, proportional reasoning, calculus etc.

An algorithm is a set of rules used to generate a sequence of actions so as to face any situation of a certain class, and reach a solution in a finite number of steps, if such a solution exists. If there is no solution, the algorithm provides the proof that there is no solution, also in a finite number of steps.

Algorithms are effective, whereas schemes are only efficient (they may even be unefficient). Effectivity means that the issue is reached in a finite number of steps.

This property is due to the binds of necessity that relate the properties of the rules of the algorithm and the properties of the objects. Schemes do not usually possess this property. The pilot might be wrong in some tricky cases, when the algorithmic procedure would have led him to a better decision.

The theoretical value of the concept of scheme comes from its generality. Gestures are organised by schemes; verbal and social interactions are organised by schemes; emotions are organised by schemes.

As a matter of fact, gestures and movements in space also address classes of situations: they always involve some geometry and some estimation of magnitudes. For instance reversing into a parking place requires non trivial estimations concerning the dimensions of the car and the parking space, the orientation of the vehicle and the front wheels, the trajectory of the vehicle at different steps, so as to decide in time when and how to move the direction of the front wheels.

Nobody has a perfect algorithm to face that kind of situation; nor have we algorithms for other gestures and movements in space like dancing in a crowded place. The organisation of gestures is not effective; yet gestures are organised by representation and rules, developed through experience and exercise.

The concept of scheme has been introduced by Kant, considerably developed by Revault d'Allonnes (1920) at the beginning of the century, then by Piaget (Piaget, 1967, for a synthesis), also by Bartlett (1932). I have tried to propose more precise and analytical definitions of this concept, with the view that it is the most essential concept of cognitive and developmental psychology (Vergnaud, 1998, 1999; Vergnaud & Récopé, 2000).

Analytical definition: *a scheme is made of four kinds of components:*
-goals and subgoals;
-rules to generate action, information seeking, and control;
-operational invariants: concepts-in-action and theorems-in-action;
-possibilities of inference.

All four components are necessary for a scheme to be a scheme. One cannot understand the organisation of activity if one forgets one of them. Goals are the intentional components, rules the generative components, operational invariants the conceptual components, inferences the components that make it possible for the subject to adjust himself to each particular situation. Most of the time, rules, operational invariants and inferences remain implicit in the behaviour of professionals. Consequently it is difficult to identify them, and easy to forget them. It is only in algorithms and some other normed professional procedures that rules are more or less made explicit. Most activities at work are performed without explicit rules. Nevertheless they exist, and it is possible to trace some of them, by observing people working and by interviewing them.

It is important to recognise that rules generate activity: not only action but also information seeking and control.

Inferences remain even more implicit, because professionals are interested in the result of their activity, not so much in the way anticipations and alternative decisions may be generated to reach that result.

Finally the organisation of activity would not be intelligible if schemes were devoid of conceptual or preconceptual representation of the objects, properties, relations, and transformations that are involved in the situation being dealt with; and if there were no relation between the properties of the subject's thought operations and the properties of the situation. These relations cannot be purely associative. It may be the case that some procedure, widely accepted in a community of practice, is not really intelligible by most users. The diagonal cross for calculating mixtures, described by Mercier in his chapter, is a good example: the lack of conceptual control is the source of many errors.

The conceptual components of schemes are essential for us to understand that activity is usually intelligent and well adapted, that different situations are handled by the same scheme, and above all that new situations, never met before, may be met with some success, owing to the decombination and recombination of existing operational invariants and rules, and the discovery of fresh ones.

A theorem-in-action is a sentence that is held to be true in action.

A concept-in-action is held to be relevant: it cannot be true or false, only relevant or irrelevant.

I will explain the difference with an example. Suppose a young person is learning farming, and tries to calculate the quantity of flour that can be made with 972 000 kg of corn. The information given is that one needs 120 kg of corn to make 100 kg of flour.

It is not that easy for learners to see what to do, even when they have learned the cross-product procedure, or the constant coefficient one, or the unit value intermediate step. Suppose that one of the participant, in a group of learners, proposes to divide 972 000 by 120. This operation corresponds to the calculation of the scalar ratio between two quantities of corn. The concept of scalar is therefore involved as a concept-in-action. But this concept alone does not make the proposal intelligible. Intelligibility comes from the possibility to use that ratio to multiply 100 kg of flour, and find the quantity of flour corresponding to 972 000 kg of corn.

$$f(972\,000) = f(a.120) \qquad \textit{find the value of } a$$
$$f(a.120) = a.f(120) = a.100$$

The last sentence is a particular case of the theorem f(a.x) = a.f(x).

This isomorphic property of the linear function is typically a theorem-in-action. Concepts-in-action (here the concept of scalar) and theorems-in-action (here the isomorphic property) are not usually made explicit.

Concepts and theorems are dialectically related, in the sense that theorems are made of concepts, and it is the function of concepts to take place in theorems. However, a theorem is never made of just one concept, and a concept is usually implied in several theorems, of different levels. This analysis is essential to understand development in education and experience, because concepts are not developed with all their properties at the same time. This analysis also enables us to understand how analogical reasoning can take place, owing to the recognition of invariants between one class of situations and another one. The idea of transfer is not a scientific concept, because it does not refer to the fundamental process of conceptualisation.

Rules and theorems-in-action are valid under certain conditions. They may be unduly applied to situations outside their scope of validity. In other words, the relevance and validity of activity is tied to the good correspondence between the characteristics of schemes and the characteristics of the situations they are supposed to address. Conditional reasoning is essential, as beautifully exemplified in the way nurses measure, evaluate and use blood-pressure (see the chapter by Noss, Hoyles and Pozzi).

In science, concepts and theorems are explicit, at least the most important ones. Words and symbols contribute to conceptualisation: symbolic invariants help in stabilising operational invariants. Scientific sentences are made of objects (also called arguments) and relational predicates (also called propositional functions, because they become propositions when the empty places are fulfilled by specified arguments). This predicative form of knowledge is of course essential; but we cannot ignore that the operational form of knowledge is even more essential, as the aim of knowledge is to enable us to act upon the world. From an epistemological point of view (not from a linguistic one) the relation between concepts-in-action and theorems-in-action in the operational form of knowledge is more or less similar to the relation between concepts and theorems in the predicative form of knowledge.

The predicative form is at the same time *more* and *less* than the operational form of knowledge:

-*more* because it is explicit and can therefore be communicated, discussed and argued;

-*less* because it is not necessarily operational, and covers only part of the knowledge we are able to use in action. This is also a characteristic of mathematics at the workplace.

The most important theoretical challenge, for researchers that try to analyse professional practices, is to trace and identify the conceptual or preconceptual components of activity, right or wrong. There are some very good examples in this book.

Another challenge, more practical in a sense, is to draw social, pedagogical, and didactical consequences from the gap between the operational forms of knowledge, as observed in the workplace (and in the class-room, quite often), and the predicative forms, as they exist in handbooks and schoolbooks. There are many connections between these two forms, and it would be a theoretical mistake to oppose one to the other, as is sometimes done to-day: when 'procedural' and 'declarative' knowledge are opposed to each other in a way that takes it for granted that 'procedural' knowledge would not be conceptual. Part of the operational form can be worded, explained, and even justified; part of the predicative form taught in the class-room is efficiently used in activity. But there is no hope that they match perfectly. One needs to transpose into the class-room some paradigmatic professional situations, not only as a way to decrease the gap, but also to make mathematics more meaningful to students. One also needs to analyse and make more explicit the mathematics used at the workplace, to identify their conditions and limits of validity.

There is a long way to go to organise initial education and in-service training in a way that would fulfil the two social functions of mathematics education: make people comfortable and happy with what they have to do at the workplace, and keep them aware that mathematics is a scientific enterprise and a cultural heritage that cannot be reduced to its use in industry and commerce.

REFERENCES

Acioly, N. M. (1994). La Juste Mesure: Une Étude des Compétences Mathématiques des Travailleurs de la Canne à Sucre du Nordeste du Brésil dans le Domaine de la Mesure. Paris 5: Universite Rene Descartes.

Acioly-Régnier, N. M. (1995). A Justa Medida: Um Estudo Sobre Competências Matemáticas de Trabalhadores da Cana de Açucar no Dominio da Medida. In Schliemann A.D. et al (Eds.), Estudos em Educaçao Matemática. Recife: Editora Universitaria da UFPE.

Bartlett, F. (1932). Remembering: A Study in Experimental and Social Psychology. London: Cambridge University Press.

Piaget, J. (1967). Biologie et Connaissance. Paris: Gallimard.

Revault d'Allonnes, G. (1920). Le Mécanisme de la Pensée: Les Schèmes Mentaux. Psychologie Française, 45(1), 161-202.

Vergnaud, G. (1996). The Theory of Conceptual Fields. In L. P. Steffe, P. Nesher, P. Cobb, G.A. Goldin and B. Greer, (Eds.), Theories of Mathematical Learning. Mahwah: Lawrence Erlbaum.

Vergnaud, G. (1998). Towards a Cognitive Theory of Practice. In A. Sierpinska and J. Kilpatrick (Eds.), Mathematics Education as a Research Domain: A Search for Identity. Dordrecht: Kluwer.

Vergnaud, G., and Récopé, M. (2000). De Revault d'Allonnes à une Théorie du Schème Aujourd'hui. Psychologie Française, 45(1), 35-50.

SECTION 1

MATHEMATICAL KNOWLEDGE IN SCHOOL AND AT WORK

SECTION 1

MATHEMATICAL KNOWLEDGE IN SCHOOL AND AT WORK

PREFACE

JEFF EVANS

The chapters in this section discuss the learning and use of mathematical ideas in a wide range of work contexts: construction, banking, nursing, flying passenger aircraft, health care, and art and design. They are written by researchers / educators in France, Germany and the UK. These authors share a dissatisfaction with traditional and 'utilitarian' views which assert that mathematical ideas and methods learned in one situation can be 'transferred' straightforwardly: that is applied in basically unaltered form in other contexts. This relates to scepticism about the pedagogies related to these approaches, for example about the idea of mathematics modules being 'bolted on' to vocational courses (Gillespie).

But these authors are also resistant to the recent 'seductive qualities' of what Evans calls 'the strong form of situated cognition'. They insist on maintaining a distinction but not a sharp disjunction between the versions of mathematics learned and used in different settings. Thus, Eberhard draws on Mercier's (1995) delineation of the theoretical and practical functions of knowledge. And Straesser contrasts the types of understandings aimed for in classroom and workplace knowledge.

Instead these chapters form part of a movement led by researchers and educators from a range of disciplines, which aims to contribute to the enhancement of peoples' opportunities to enrich their vocational and other everyday practice, by drawing on learnings from school, college and university settings. Otherwise put, to reformulate the 'problem of transfer'. Though their approaches differ somewhat, they share certain metaphors for describing their aims as analysing similarities and differences across the 'boundaries' between settings and the practices and relations based therein (Noss, Hoyles and Pozzi; Evans), and to 'build bridges' (Straesser). Noss et al. recommend 'exploring the semantics of mathematics in different ... settings' so as to 'avoid conflict between mathematics as an object of study and mathematics in use, but, rather, to see them in dialectical relationship'. Thus their notion of situated abstraction aims to capture the way that the mathematical meanings developed by learners are 'shaped and constrained by tools and language', as well as 'some

Bessot & Ridgway (eds.), Education for Mathematics in the Workplace, 3—4.

essence of the salient mathematical relationships'. Evans's approach emphasises rather more the shaping of practice by discourse, and shows that it is possible for school / college mathematics to aim at building bridges by charting the flow of meaning through key signifiers -- though such flows are often unpredictable.

One crucial, but problematical, concept, is that of the 'visibility' of mathematics in different settings, and in different tools, such as computer software (Noss et al., Straesser, Gillespie). Mentioning visibility begs the question: By whom? Variations across groups and individuals arise from (at least) two sources. First, different trainings are, to a great extent, trainings in 'different visions' i.e. to see different things. And, second, different contexts require attention to seeing different things. For this reason, there are quite understandable difficulties for students in recognising an opportunity to use (Bernstein, 1996; Cooper and Dunne, 2000) a particular 'mathematical' technique (Eberhard).

In the light of these arguments, perhaps the notion of 'visibility ' of a particular mathematical idea or method needs to be replaced by the notion of the participants' vision supported by a particular discursive practice.

A range of methodologies are used by the researchers here:
-documents (e.g. professional textbooks);
-ethnographic observation;
-semi-structured interviews;
-task-based interviews;
-and observation of learning and teaching.

Besides attending to routine flows of activity, or interview accounts of these, Noss et al. also point to the importance of studying 'breakdowns' of routine activity.

Several authors provide suggestions for teaching methods aimed at building bridges between school or college learning and workplace applications: Gillespie's chapter is particularly rich in examples. Suggestions for further research are given in particular chapters and also in Straesser's concluding chapter to the book.

To summarise, the authors in this section collaborate in pointing to the need for a re-thinking of the traditional approach to the transfer of school or college-mathematics to the workplace. What is needed instead is an approach to what might be called the 'translation' across practices, and across the different 'perspectives' of vocational and mathematics specialists – so as to connect mathematical theory to professional practice.

REFERENCES

Bernstein, B. (1996). *Pedagogy, Symbolic Control and Identity*. London: Taylor and Francis.

Cooper, B., and Dunne, M. (2000). *Assessing Children's Mathematical Knowledge: Social Class, Sex and Problem-solving*. Buckingham: Open University Press.

Mercier, A. (1995). Le Traitement Public d'Elements Prives du Rapport des Élèves aux Objets de Savoir Mathématiques. In G. Arsac, et al (Eds.), *Differents Types de Savoir et Leur Articulation*. Grenoble: La Pensee Sauvage.

CHAPTER 1

THE TRANSFER OF MATHEMATICS LEARNING FROM SCHOOL TO WORK NOT STRAIGHTFORWARD BUT NOT IMPOSSIBLE EITHER!

JEFF EVANS

Abstract. The take-up of opportunities for applying school learning is often disappointing - to teachers, parents, employers, and many pupils. Not surprisingly, there is much controversy among researchers in mathematics education and related fields, as to the reasons. Here I argue that neither traditional views, with their simplistic faith in the basic continuity of knowledge across contexts, nor currently popular 'insulationist' views such as the strong form of situated cognition, which claims that transfer is basically not possible, are adequate. Instead, giving examples of why transfer is not dependable and often very difficult, I argue that, in order to realise any possibilities of transfer, it is necessary to analyse the discourses involved as systems of signs, and to look for appropriate points of articulation between them. I aim thereby to clarify the problem and to contribute to efforts to help learners to build bridges between different practices, particularly between school and work.

1. INTRODUCTION

If schooling is to be relevant to settings and activities outside itself, then we need some account of how learning from the school can be applied in, or 'transferred' to, other contexts. The *transfer* of learning can be considered to refer in general to the use of ideas and knowledge from one context in another. This might involve:

(i) the reformulation of academic discourses as school subjects;

(ii) the use of a school subject like mathematics outside of its own domain, e.g. in physics or economics;

(iii) the application of knowledge from pedagogic contexts to work or everyday activities; or

(iv) the 'harnessing' of out-of-school activities in the teaching of school subjects.

This is clearly an especially important set of issues for mathematics: it is claimed to have very wide applicability across the curriculum, and outside the school. Here I am especially interested in issues around (iii), and, in particular, applications in work contexts.

However, in practice, transfer remains a difficult problem: one cannot *depend* on its being accomplished, by a particular learner, in a particular situation. There is

Bessot & Ridgway (eds.), Education for Mathematics in the Workplace, 5—15.

much anecdotal evidence that teaching often has disappointing results in this respect, or put another way, that students often 'fail' to accomplish transfer. And recent research has shown striking differences between levels of performance - and methods used - in work, or everyday situations, on the one hand, and in school, or school-type tasks, on the other (e.g. Lave, 1988; Nunes, Schliemann and Carraher, 1993).

2. VIEWS ON THE TRANSFER OF LEARNING IN MATHEMATICS

Indeed, questions around transfer are strongly contested, and a variety of views proliferate in educational circles, as well as in psychology and sociology. The discussion has been especially vibrant in mathematics education in the last 10 or 15 years. Here *traditional* approaches include views favouring the use of behavioural learning objectives, 'basic skills' approaches (Eraut, 1996), and 'utilitarian' views such as those of the Cockcroft Report (1982). They share several important ideas. Learning is seen as involving the transmission and internalisation of a body of knowledge (and skills). A problem or 'task', and the mathematical thinking involved in addressing it, are seen as able to be described in abstract, e.g. as 'taking a percentage' or 'proportional reasoning', and thus as separated from the context. Some tasks are then seen as *essentially* mathematical, and hence it is claimed to be possible to talk about 'the same mathematical task' occurring across several different contexts. And contexts themselves are described 'naturally', as school mathematics or 'business maths', say, apparently not needing any further description or analysis. Therefore traditional views expect that the 'transfer of learning', e.g. from school to everyday situations, should be relatively unproblematical - at least, in principle, for those who have been properly taught.

In recent years, several strongly sceptical arguments have emerged, and been widely discussed, based on ideas of 'insulation' (Muller and Taylor, 1995) between contexts and forms of knowledge. One example is the *strong form* of situated cognition: a major text for this view has been Jean Lave's *Cognition in Practice* (1988). It argues that there is a *disjunction* 3between doing mathematics problems in school, and numerate problems in everyday life, as these different contexts are characterised by different *structuring resources* - including ongoing activities, social relationships, cultural forms of quantity such as money. These worlds and the practices in play in them are *disjoint,* and subjects' thinking is specific to these practices, and these settings. Thus transfer of learning from school / academic contexts to outside ones is judged to be pretty hopeless.

Jean Lave's more recent work (Chaiklin and Lave, 1993; Lave, 1996) is no longer so concerned to stress disjunctions between practices. Her current position acknowledges that no practice could ever be completely closed. Her approach consists of studying learning within communities of practice, and the bridges between them; in particular, the social relations, and identities across them. However, in this approach, there seems again to be an assumption that practices and communities of practice can be seen as 'natural' - whereas I argue that they must be described and analysed as *socially constructed.*

This brief discussion (see also Evans, 1999, 2000) suggests a need for a reformulation of the problem of transfer. There are a number of issues or gaps in earlier accounts that need attention, including:

(1) how to define and delineate the practices in question - namely, both school or college mathematics and 'target practices' for transfer - and related contexts of thinking, activity and learning;

(2) how to describe the relations between practices and communities of practice, e.g. what the boundaries or bridges between them might be like;

(3) how to acknowledge the inter-relationships of thought, and feeling;

(4) how to design pedagogic practices that will facilitate transfer ('teaching for transfer').

Currently, a number of areas, besides mathematics education, are contributing to relevant discussions, and seem to me to be converging in their analyses. These include: developmental psychology - e.g. Nunes et al. (1993), Saxe (1991); cognitive psychology - e.g. Singley and Anderson (1989), Anderson et al. (1996), Pea (1987); and approaches drawing on discourse theory (including poststructuralist insights) - e.g. Walkerdine (1988); Walkerdine and Girls & Maths Unit (1989); Muller and Taylor (1995); Evans and Tsatsaroni (1994, 1996); Evans (1999, 2000).

3. CONCEPTUALISING BOUNDARIES AND BRIDGES

(1) defining and delineating the practices in question, and their related contexts of thinking, activity and learning

The approach I am proposing focusses on *practices*: examples would be school mathematics, research mathematics, work practices such as nursing and banking, apprenticeship e.g. into tailoring, and everyday practices such as shopping. Each practice is constituted by *discourses*, systems of ideas expressed in terms of *signifiers* and *signifieds*, based in relations of similarity and difference. These discourses give meaning to the practice by expressing the *goals* and *values* of the practice, and *regulate* it in a systematic way, by setting down standards of performance. Important practices are associated with a community of practice, a subculture of individuals with (some) shared goals, and a set of social relations (power, difference) with different members of the community taking up different *subject-positions*. For example, the basic positions available in school mathematics are normally 'teacher' and 'pupil'; in shopping or street-selling, they would be 'seller' and 'buyer'. In a particular setting, we can analyse the practices *at play*, that would be involved in the positioning of participants.

This approach, like situated cognition, recognises different practices as in principle distinct, as discontinuous - e.g. school mathematics and everyday practices like street selling. But, using the approach recommended here, we can go further. Language and meaning in the discourses involved can be analysed by considering relations of signification - relations of similarity and difference between signifiers (words, gestures, sounds, etc.) and signifieds (conceptions) - and devices such as metaphor and metonymy. So far this draws on Saussure's structural linguistics.

Going further, Walkerdine and others have shown how to use poststructuralist ideas about the inevitable tendency of the signifier to slip into other contexts, thereby making links with other discourses, and producing a play of multiple meanings, so as to provide insight into meaning-making in mathematics; see e.g. Walkerdine (1988, Ch.2) on children's use of language to indicate relations of size, Brown (1994) and Evans and Tsatsaroni (1994). Thus, rather than attempting to specify the context of a school mathematics problem by looking only at its wording, or by naming the context as if simply based in 'natural' settings, we can describe it as socially constructed in discourse - through attention to particular signifiers and their relations, e.g. in interview transcripts and other texts.

(2) describing the relations between practices

Examining the relations among signifiers and signifieds enables us to analyse the *similarities and differences* between discourses, for example between school and everyday maths. This is in turn necessary for understanding possible bridges between practices.

Contrary to the hopelessness of the strong form of situated cognition, I argue that it is possible to build bridges between practices, by trying to identify areas where out-of-school practices might usefully 'overlap' or inter-relate with school or college mathematics. This requires first of all that distinctions are made between those relations of signification in the learner's everyday practices that provide *fruitful* 'points of articulation' with school or college mathematics, and those that may be *misleading*. An example of a misleading inter-relation would be the problematical attempt to harness the use of 'more' in the home - where its opposite is *no more* (as in 'no more ice cream for you') - to help teach 'more' vs. *less* as an oppositional couple at school. The pupils are likely to be confused because what appears to be 'the same' signifier has a different meaning (signified) in the home and the school discourses (Walkerdine and Girls & Mathematics Unit, 1989: 52-53).

Once potentially fruitful points of articulation between school or college mathematics and the target practices are identified, then pedagogic strategies can be brought into play; see the next section.

(3) acknowledging the inter-relationships of thought and feeling

Many accounts of cognition generally, and even situated cognition, largely ignore the area of emotion. Taylor has discussed the power of desire at the individual and the societal level (1989). Walkerdine has emphasised the importance of the relations between cognition and affect: 'meanings are not just intellectual' (Walkerdine & Girls and Maths Unit, 1989, p.52). Affect can usefully be seen as a charge attached to particular signifiers that make up *chains of signification*. This charge can flow from one signifier to another, by *displacement*. Indeed it can be argued that insights from psychoanalysis can allow us a fuller consideration of the affective (Walkerdine, 1988; Evans and Tsatsaroni, 1994, 1996).

The allowing in of affect or emotion at the level of the particular subject is not just an 'optional extra'. Other work shows that whenever a teacher reaches outside of mathematics for an example as illustration, the mathematics is 'at risk'; for example, when illustrating mathematics in the context of shopping with 'Mummy', when the mother 'has financial difficulties, ... is sick far away or deceased' (Adda, 1986, p.59). This is because of the fundamental character of language, its ability to produce 'multiple meanings': the signifier can always break with any given context, and be inscribed in new contexts without limit (Evans and Tsatsaroni, 1994, p.184).

Thus another reason that a particular set of relations of signification may not be an apt source of fruitful articulations, in attempting to *harness* - everyday life for school purposes, is that these relations may be *distressing* or *distracting* - and not only misleading. Similar issues may well arise in attempts to *transfer* from school learning to work; see the case study below and further examples in Evans (2000).

4. TEACHING FOR TRANSFER

(4) designing pedagogic practices that will facilitate transfer

Besides attending to fruitful (non-misleading, non-distracting) points of articulation, we must structure the pedagogic discourse so as to work systematically through a process of *translation,* a kind of 'meaning-making'. This involves the *rearticulation* of signifiers and signifieds linked in one set of signs, and their reinsertion into a new set of signs. This is done through the construction of chains of meaning (Walkerdine, 1988, p.128ff), as shown below.

A very simple example of what I mean is that of a mother teaching a child to count by the following transformations. This illustrates how transfer from home to school practices might be accomplished by a careful sequence of steps (see also Muller and Taylor, 1995):

Step

1 Particular Child *(signified)*

Name of Child *(signifier)*

2 Name of Child *(signified)*

Finger *(iconic signifier)*

3 Finger *(signified)*

Spoken Numeral *(symbolic signifier)*

4 Spoken Numeral *(signified)*

Written Numeral *(symbolic signifier)*

At each step, the mother-teacher, encouraged the child to form a sign linking a new signifier (word, gesture) to a signified, which had been the signifier at the previous stage. Each step thereby creates a new set of signs. The chain of meaning moves as follows: actual child (more precisely, *the idea* of the child) -name of child -iconic signifier -spoken symbolic signifier -written symbolic signifier.

Here, the different steps do not really represent different discourses, but they nevertheless show how a set of carefully constructed links between signifier and signified could provide the bridges for crossing boundaries between home and school discourses, or for 'teaching for transfer'. Another example is provided by a teacher teaching children to add in Walkerdine (1988, pp.122 ff.).

Of course, transfer can go wrong if the issues above are neglected; an example is provided by a primary school 'shopping game' (Walkerdine, 1988, Ch. 7). There a boy made 'errors' in his sums because he did not realise that, in the game, one was allowed - indeed, one was *required* by the rules, made to ensure the game's pedagogic effectiveness - to start afresh with a new 10p after each purchase. Though the child *called up* - that is, identified the task as - practical shopping, through which he 'made sense' of the apparent demands of the task, he nonetheless made errors because he was *positioned in*, and *regulated by*, the pedagogic shopping game.

While some aspects of everyday shopping practice were also useful in the game, such as remembering the familiar result that 'when you have 10p and buy something worth 9p, you will have 1p left', other aspects of shopping - for example, the knowledge that one must give up money to obtain the purchase - were not 'included' in the discourse of the school shopping game. Also, importantly, the goals and purposes were quite different in the two practices.

Thus, Walkerdine argues that activity within one discourse, e.g. playing a particular card game, will help with – that is, can be 'harnessed' for - school mathematics in those, and only those, aspects of the game which are both contained in school mathematics and which enter into similar relations of signification (Walkerdine, 1988, pp. 115 ff.).

Schliemann (1995), in a paper concerned with the viability of harnessing mathematics from everyday settings to help with learning school maths, reaches a similar conclusion:

> ... mathematical knowledge developed in everyday contexts is flexible and general. Strategies developed to solve problems in a specific context can be applied to other contexts, *provided that the relations between the quantities in the target context are known by the subject as being related in the same manner as the quantities in the initial context are.* (p.49, my emphasis)

These views, from developmental psychologists, seem to converge with those of some cognitive psychologists, who argue that transfer between tasks is a function of the degree to which the tasks share symbolic components (Singley and Anderson, 1989; Anderson et al., 1996). In order to understand fully the possibilities for translation, I argue that it is necessary to consider *both similarities and differences* across practices (Evans, 1999, 2000). For an approach sharing this emphasis, see Noss, Hoyles and Pozzi (this volume).

Let us take as an example the teaching of methods of mathematical / statistical modelling in a college (or school) Business Studies course. It is an aim of the course that, during their mid-course placement in a local business, the students will be able to recognise situations - such as variations in sales across companies or time periods - that might usefully be modelled mathematically (to simplify: using one particular model, multiple regression) and to produce a basic model in this work situation.

Here I can present only a sketch of some useful guidelines for teaching / learning for transfer, which have been sifted from earlier writers (Anderson et al., 1996; Pea, 1987), and developed from the analysis given above:

(a) Show how to perform a detailed analysis of the shared or similar components - *and the different aspects* - of the initial and target tasks. In the illustration given here:

differences: business - isolation of 'variables' from an analysis of a problem, as described in the discourse of the business situation;

college maths and stats. - variables (tend to be) pre-conceptualised and listed.

similarities: e.g. distinguishing of target outcome ('dependent variable') and possible influences ('predictor variables').

Note, however, the differences in language in the two discourses, business and 'college maths and stats.'.

(b) Include the ability to transfer as a specific, and explicit, goal. This means establishing links between the two situations, and the related discourses and practices, by *translating* between the terms / languages used, and by *generalising* the methods used across contexts (see (d)).

(c) In teaching the initial task, seek to incorporate a balance of generality and situational features. That is, anticipate what will come to be seen as *similarities* across situations, and *differences*, respectively.

(d) Teach the initial task in more than one context. For example, use examples of the determinants of salaries within a particular professional group (by number of years of service, educational qualifications, etc.).

(e) Allow practice in recognising the cues that signal the relevance or 'applicability' of an available skill. In mathematics these cues are recurrent features of pattern, structure, or relationship[1]. For example, in the above, both situations have been analysed (and simplified by *assumptions*) to depict a quadratic relationship between two variables (that holds independently of place and time). This is the sort of analysis we would hope our students could do as the basis of mathematical, statistical or business economics modelling.

(f) Allow repetition or practice on the target task .. This will help the student to appreciate the possible range of generalisation, and the constraints on it, resulting from crucial differences in discourses.

For further discussion of the problem of transfer in mathematics teaching, see e.g. Boaler (1997) and Masingila (1996).

5. IMPLICATIONS FOR RESEARCH

In the discussion above, I pointed to the need to specify the context of a school mathematics problem by:

- analysing the practices at play in the setting; and
- attending to particular signifiers and their relations of similarity and difference, as they are used in social interaction.

Thus some of our research effort in this area can usefully be involved in producing and reading interview transcripts, e.g. of learners or workers engaged in problem-solving.

Finally, as an illustration of sensitivity to ways of promoting transfer - but also to the barriers and pitfalls - I include a brief reference to a case study from one of a series of my own research interviews (Evans, 1999, 2000): the responses of 'Donald' to one of the problems presented to a sample of social science undergraduates for solving (concerning a graph showing how the price of gold varied over one day's trading in London; see Figure 1.).

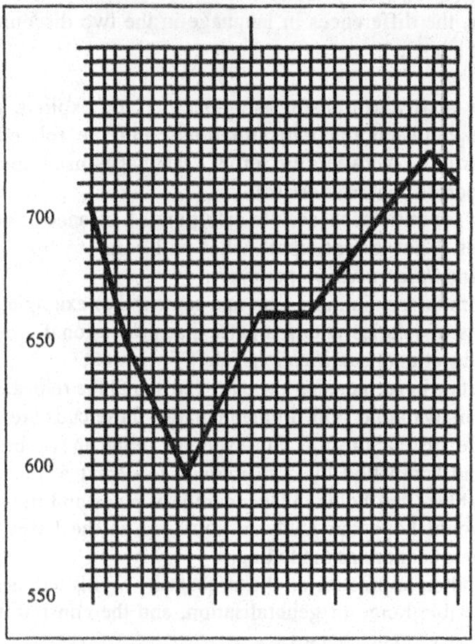

Figure 1. Interview Problem on Changes in the Price of Gold[2]
(The London Gold Price - January 23rd 1980)
Source: Evans (2000), based on Sewell (1981)

This case study shows several things:

(a) Donald is apparently able to focus on *discursive similarities and differences*: he seems able to read the diagram as a 'chart' (business maths) or as a 'graph' (college maths), and to recognise the connections between a 'trend' and a 'gradient' (respectively).

(b) He is also aware of the different goals of the two practices, that relate to different objectives in using the graph. In *business*, the objectives are implicitly competitive, to make comparisons across persons or groups, and are growth-orientated, to make comparisons over time; in *college mathematics*, the objectives are less comparative, aimed at analysing the qualities of the curve, including the rate of change. He is aware of *different values and standards of regulation*, in particular of *precision*, required in the two discourses of college mathematics, and 'business maths'.

(c) He is also open about the *different feelings* evoked by the two practices. For example, his awareness of the different goals of the two practices (see (b) above) is sometimes painful (Evans, 2000).

(d) He shows himself able and willing to use both college mathematics and money-market maths; further it appears he is able to choose which practice to use to address the problem in the interview, to decide whether to apply his (more precise) college maths methods of calculating gradients to the problem of deciding when the relevant price was rising faster. Though not certain, it appears that Donald is able to bridge the two practices, i.e. to transfer his college maths methods to deal with a problem involving charts (assuming he were convinced of the need) on the basis, perhaps, of some idea of *economy of cognitive effort* (cf. Pea, 1987).

6. CONCLUSION AND DIRECTIONS FOR RESEARCH

Here I set down a general set of conclusions concerning the transfer - or translation - of learning. I am proposing this re-labelling, because of the links between the notion of transfer and the traditional views criticised above, as well as widespread dissatisfaction with the notion (e.g. Lave, 1988, 1996).

Continuities between practices (e.g. school and out-of-school activities) are not as straightforward as traditional views assume, and hence scepticism is in order about the idea that transfer of learning between practices is itself straightforward.

We can accept the views of researchers in situated cognition and others that there is a *distinction* - but not necessarily a *disjunction* - between doing mathematics problems in educational settings, and numerate problems in everyday life. Thus we can acknowledge that transfer is not dependable and often difficult. But it is not impossible, and hence we can be more optimistic than these other approaches suggest.

In teaching and learning, bridges between practices can be built, by (a) describing the practices involved (in the transfer relationship), and analysing the related discourses as systems of signs; and by (b) analysing the *similarities and differences* between discourses (e.g. college vs. everyday mathematics), so as to identify fruitful 'points of articulation' between mathematics in educational settings and outside ('target') activities.

The inter-relationships of thought and feeling have received insufficient emphasis in most earlier discussions of transfer. They must be allowed for, because of the risks of attempts to inter-relate practices in ways that may distract the learner, and because of the involvement of emotion in the 'flow of language'.

The inevitable tendency of language to flow in unexpected ways and generally to assume multiple meanings within different practices, constitutes a severe *limitation* on the possibilities of any intended transfer. Yet this ability of a signifier to form different signs also provides the *basis* for any transfer possibilities. Thus, although the successful crossing of bridges cannot be guaranteed 'risk-free', this paper has sketched some steps it is *necessary* to follow. For anything like transfer to occur, a 'translation' across discourses would have to be accomplished through careful attention to the relating of signifiers and signifieds in particular chains of meaning. This translation is not straightforward, but it often will be possible.

Ways of designing pedagogic practices can be developed that will facilitate transfer, including: task analyses of similarities and differences between discourses; incorporating a balance of generality and situational features in teaching the initial task; and providing practice in recognising cues for the applicability of a specific idea or method.

To develop the study of transfer from school to work, we need research programmes including a focus on sign systems and meanings, as outlined above, and more widespread workplace studies in the styles of Recife (e.g. Nunes *et al.*, 1993) and the London Institute (e.g. Noss & Hoyles, 1996, this volume).

Acknowledgement

I thank participants in the ESRC-sponsored seminar series, Producing a Public Understanding of Mathematics, held at the University of Birmingham, 1998-99, for comments on an earlier draft of this paper.

7. REFERENCES

Adda, J. (1986). Fight Against Academic Failure in Mathematics. In P. Damerow, et al (Eds.), *Mathematics for All* (pp. 58-61). Paris: UNESCO.

Anderson, J. R., Reder, L.M. , and Simon, H. A. (1996). Situated Learning and Education. *Educational Researcher, 25*(4), 5-11.

Bernstein, B. (1996). *Pedagogy, Symbolic Control and Identity.* London: Taylor and Francis.

Boaler, J. (1997). *Experiencing School Mathematics.* Milton Keynes: Open University Press.

Brown, T. (1994). A Post-structuralist Account of Mathematical Learning. In P. Ernest (Ed.), *Mathematics, Education and Philosophy: An International Perspective* (pp. 154-161). London: Falmer Press.

Chaiklin, S., and Lave, J. (1993). *Understanding Practice: Perspectives on Activity and Context.* Cambridge: Cambridge University Press.

Eraut, M. (1996). *The New Discourse of Vocational Education and Training: A Framework for Clarifying Assumptions, Challenging the Rhetoric and Planning Useful, Theoretically Informed Research.* Paper presented at the European Conference in Educational Research, Seville.

Evans, J. (1999). Building Bridges: Reflections on the Problem of Transfer of Learning in Mathematics. *Educational Studies in Mathematics, Special Issue on the Contexts of Teaching and Learning Mathematics* (39), 23-44.

Evans, J. (2000). *Adults' Mathematical Thinking and Emotions: A Study of Numerate Practices.* London: Falmer Press.

Evans, J., and Tsatsaroni, A. (1994). Language and Subjectivity in the Mathematics Classroom. In S. Lerman (Ed.), *The Culture of the Mathematics Classroom* (pp. 169-190). Dordrecht: Kluwer.

Evans, J., and Tsatsaroni, A. (1996). Linking the Cognitive and the Affective in Educational Research: Cognitivist, Psychoanalytic and Poststructuralist Models. *British Educational Research Journal, Special Issue on Poststructuralism and Postmodernism* (21(3)), 347-358.

Lave, J. (1988). *Cognition in Practice: Mind, Mathematics and Culture in Everyday Life*. Cambridge: Cambridge University Press.

Lave, J. (1996). Teaching as Learning. *Practice, Mind, Culture Activity, 3*(3), 149-164.

Masingila, J. A., Davidenko, S. and Prus-Wisniowska, E. (1996). Mathematics Learning and Practice In and Out of School: a Framework for Connecting these Experiences. *Educational Studies in Mathematics, 31*, 175-200.

Muller, J., and Taylor, N. (1995). Schooling and Everyday Life: Knowledges Sacred and Profane. *Social Epistemology, 9*(3), 257-275.

Noss, R., and Hoyles, C. (1996). *Windows on Mathematical Meanings: Learning Cultures and Computers*. Dordrecht: Kluwer.

Nunes, T., Schliemann, A., and Carraher, D. (1993). *Street Mathematics and School Mathematics*. Cambridge: Cambridge University Press.

Pea, R. (1987). Socialising the Knowledge Transfer Problem. *International Journal of Educational Research, 11* (Special Issue on Acquisition and Transfer of Knowledge and Cognitive Skill).

Saxe, G. (1991). *Culture and Cognitive Development: Studies in Mathematical Understanding*. Hillsdale, NJ: Lawrence Erlbaum Associates.

Schliemann, A. (1995). *Some Concerns about Bringing Everyday Mathematics to Mathematics Education*. Paper presented at the 19th International Conference for the Psychology of Mathematics Education (PME-19), Recife, Brasil.

Sewell, B. (1981). *Use of Mathematics by Adults in Everyday Life*. Leicester: Advisory Council for Adult and Continuing Education.

Singley, J. K., and Anderson, J. R. (1989). *The Transfer of Cognitive Skill*. London: Harvard University Press.

Taylor, N. (1989). Let Them Eat Cake. In C. Keitel, A. Bishop, P. Damerow and P. Gerdes, P. (Eds.), *Mathematics, Education and Society*. Paris: UNESCO (Proceedings of Day 5, ICME-6).

Walkerdine, V. (1988). *The Mastery of Reason: Cognitive Development and the Production of Rationality*. London: Routledge Kegan Paul.

Walkerdine, V. (1989). *Girls and Mathematics Unit. Counting Girls Out*. London: Virago.

8. NOTES

[1] These recurrent features come under what Nunes et al. (1993), following Gérard Vergnaud, call 'invariants'

[2] The problem read as follows: this graph shows how the price of gold (in dollars per fine ounce) varied during one day's trading in London. Which part of the graph shows where the price was rising fastest? What was the lowest price that day?

CHAPTER 2

WORKING KNOWLEDGE: MATHEMATICS IN USE

RICHARD NOSS, CELIA HOYLES AND STEFANO POZZI

Abstract. In this chapter, we explore how mathematics is used and described in work in the context of investment bank employees, paediatric nurses and commercial pilots. We distinguish the visible mathematics of the workplace, that is the mathematics referenced in textbooks and by experts, from the mathematics observed in ethnographic observations of the practice. Our conclusions are that practitioners do use mathematics in their work, but what they use and how they use it may not be predictable from considerations of general mathematical methods. Strategies at work depend on whether or not the activity is routine and on the material resources at hand.

For the past 15 years, studies of adults' behaviour in the workplace have had an important impact on the way we think about mathematical reasoning. For example, research with dairy workers (Scribner, 1986), carpenters (Millroy, 1992), carpet-layers (Masingila, 1994), seamstresses (Harris, 1987, Hancock, 1996), automotive industry workers (Smith & Douglas, 1997) and civil engineers (Hall & Stevens, 1995, 1996) have all pointed towards a similar conclusion: that most adults use mathematics to make sense of situations in ways which differ quite radically from those of mathematicians (for some fascinating material concerning the ways in which various professional groups, especially architects, incorporate mathematics into their practice, see Barrallo, 1998). Rather than striving towards consistency and generality, problem solving at work is characterised by a pragmatic agenda and geared to solving particular problems. Occupational or professional concerns take precedence over those that are mathematical. This body of work has provided important insights into how people conceptualise the role of mathematics in their work. What emerges clearly is that people develop strategies to carry out their work quickly and efficiently using their knowledge and experience but also by exploiting features of the environment and its local regularities. It is in this way that knowledge both shapes and is shaped by workplace activity.

One strong theoretical orientation in studies of the workplace can be broadly grouped under the heading of situated cognition (see, for example, Lave, 1988; Lave & Wenger, 1991). Partly based on strands within Activity theory, situated cognition provides a view of intellectual work (including mathematical work) which is inseparable from its socio–cultural contexts. Indeed, Activity theory itself has had an increasing influence on the way researchers conceptualise mathematical

Bessot & Ridgway (eds.), Education for Mathematics in the Workplace, 17—35.

understanding, through its emphasis on how acts of problem solving are contingent upon structuring resources such as notational systems, physical and computational tools, and work protocols (Gagliardi, 1990). These artifacts are described by Leont'ev (1978) as 'crystallised operations', borne out of the needs within a given set of social practices, and in turn playing their part in shaping and restructuring future practices. From this perspective, mathematical activity in work is deeply intertwined with the complexities of working practices, and gives credence to the notion that we cannot develop a theory of mathematical competence without developing a theory of working knowledge.

There has been another strand of research around the theme of mathematics in work which has been more policy-oriented, attempting to drive curriculum reform in schools through surveys of 'the mathematics needed' in different workplaces (from the UK see, for example, Fitzgerald, Purdy & Rich, 1981; SCAA, 1997). Arising from a utilitarian agenda, this research has typically treated as unproblematic the notion that mathematics education can deliver a general intellectual resource which can be transferred from the classroom to the workplace (for a critique of such utilitarian approaches, see Noss, 1998). Lacking a clear epistemological focus, such studies are beset with methodological problems, not least since the subjects of these studies rarely describe their activities in mathematical terms and often declare that they use little if any mathematics in their work (for more detailed accounts, see Wolf, 1984; Harris, 1991). Clearly, attempting crude behavioural classifications based only on the mathematics of school curricula fails to evoke the authentic details of real work practices, unlike the ethnographies provided by the studies in the situated paradigm described earlier.

However, we believe there is a fundamental difficulty that often characterises studies in both traditions which relates to the conceptualisation of mathematics. As Nunes (1992) has pointed out: if mathematics is seen as an arbitrary set of symbolic representations and procedures, then the definition of mathematical knowledge remains implicit and based on culturally-defined notions of content (e.g. number and calculation, geometrical patterns, algebra). This view inevitably emphasises the variation in mathematics across boundaries — between the cultures of school and work, between different communities. In consequence, similarities are often ignored, resulting in a limited view of mathematical knowledge that fails to recognise what is epistemologically invariant under such cultural transformations. Policy studies focus only on the most 'visible' mathematics in the workplace based on the content and conventions of school mathematics, which can only confirm long-established curriculum content such as numeracy, mensuration and basic graphical interpretation. Within situated cognition research on the other hand, mathematics can only be characterised as fragments in a setting taking for granted the supposed impossibility of conceptualising mathematics in terms of more general strategies. Nunes suggests that a more fruitful way to investigate mathematics in use would be to focus on the invariants of mathematical structures, arguing as follows:

> In this view, 'mathematizing' reality is representing reality in such a way that (a) more knowledge about the represented reality can be generated through inferences using mental representations, and (b) there is no need to manipulate reality further in order to

> verify this new knowledge. Invariant logical structures are embedded in mathematical knowledge, regardless of whether mathematical knowledge is developed in or out of school. It is the ability to make inferences using these structures - not the content of knowledge - that distinguishes mathematical from other kinds of knowledge (Nunes 1992, p. 558).

An emphasis on mathematical structures allows researchers to seek out both similarities and differences across boundaries, and this is the agenda we have tried to follow. We have for some time been exploring the semantics of mathematics in different classroom settings, in an attempt to problematise common dichotomies, such as formal-informal, concrete-abstract, contextualised-decontextualised. By emphasising questions of meanings and their construction, we have attempted to avoid conflict between mathematics as an object of study and mathematics in use, but rather to see them in dialectical relationship. In Noss and Hoyles (1996a), we coined the notion of *situated abstraction* to catch how the mathematical meanings children develop are shaped and constrained by tools and language, yet simultaneously capture some essence of the salient mathematical relationships. We want now to extend this notion beyond the confines of its origins in computer environments and to think further about what exactly is situated about a situated abstraction — how does the setting act back on conceptions? Researching in the context of the workplace provided us with just the right context for furthering this agenda. It also afforded us the opportunity of making connections between what are largely two disparate worlds — the world of mathematics learning and the world of mathematics in work. We base the remainder of this chapter on the findings of a recent project[1] where we explored how mathematics was used in various work contexts in order to tease out its potential and limitations in terms of developing useful models of work practices.

1. THE STUDY

Our project consisted of a series of three studies involving investment bank employees, paediatric nurses in a children's hospital and commercial pilots. These groups are similar in three ways: they have comparable mathematical entry requirements instantiated in specific training programmes, and they engage with mathematics in 'error-critical' activity — there is little or no room for error in for example, the confirmation of financial transactions, drug administration or landing an aircraft. Our initial aims were exploratory: we wanted to examine the relationship between practical, professional and mathematical knowledge, and how these resources were used together or separately at work. Our approach therefore involved a number of methodological gambits starting with what we could see as the mathematisable elements of practice i.e. elements which, from a mathematical perspective, seemed amenable to mathematical modelling in some form.

We elaborated our initial objectives by formulating a series of questions addressed over two stages of each study.

Stage 1: Developing a map of mathematical workplace activities

-to what extent did workplace activities incorporate mathematical elements?

-how do these mathematical elements relate to practitioner knowledge?

-how do tools and technologies shape the relationship between mathematical elements and practitioner knowledge?

Stage 2: Mathematising workplace activities

-what workplace activities are amenable to mathematisation?

-how do practitioners respond to progressive mathematisation of their workplace activities?

-can practitioners make sense of mathematical models of their workplace activities and are they able to construct models for themselves?

-how does the analysis of practitioners' mathematisations illuminate connections between mathematics and practitioner knowledge?

Our general methodological approach was based on an iterative process of data collection, analysis and hypothesis generation, and involved examining the mathematics of the workplace (Stage 1) and studying the process of mathematisation (Stage 2). Stage 1 methods comprised documentary analysis, interviews with senior staff in each profession, initial interviews with practitioner volunteers, and ethnographic observation of these subjects in the workplace. In Stage 2, we undertook task-based interviews with a small group of practitioners, questionnaires with a larger group, along with observation of their exploration and construction of computer-based models within a number of teaching experiments we devised.

1.1. Documentary Analysis and Interviews

For each study, we analysed a range of teaching texts aimed at entrants to the profession to develop a preliminary audit of the mathematics expected of practitioners. Two criteria were used in choosing texts: that they were at approximately comparable mathematical levels, and that they were either written for the purpose of supporting workplace mathematics or included substantial sections of mathematical content[2]. The analysis involved listing mathematical elements and, where possible, identifying workplace activities in which these elements would be used.

Semi-structured interviews were conducted with senior staff in each profession, to validate the accuracy of the documentary analysis, provide further contextual information on activities which — from their point of view — involved mathematics, and establish protocols for observations in the workplace. Senior staff also assisted in recruiting volunteer subjects for the subsequent stages of the studies. Fieldnotes from interviews were used to generate more authentic and contextualised audits of relevant activities, and to plan appropriate protocols to solicit information from subjects during pre-interviews and ethnographic observation.

Semi-structured interviews were also conducted with volunteer subjects. These focussed on professional and mathematical issues, including requests for descriptions of 'typical days', workplace activities considered to involve mathematics, and attitudes to mathematics in school, training and at work. Interview notes and transcripts provided further evaluation and contextualisation of the previously audited activities, as well as identifying changes in practitioners' attitudes to mathematics from school to work.

1.2. Ethnographic observation

Ethnographic observation involved shadowing and interviewing subjects informally at work: in the bank, on the ward and in the cockpit. Individual sessions lasted for 1-3 hours, with each subject visited several times. Fieldnotes and, where possible, recordings of discussions between subjects and their colleagues were taken, and copies of artifacts and charts collected. In the banking study, it was only possible to take fieldnotes, although these did include copies of the resources used. In the nursing study, audio recordings of discussions were made on the ward. In the flying study, video recordings were made in the cockpit, which captured data of discussion and non-verbal communication as well as the resources and instrumentation used.

Fieldnotes, copies of resources, transcripts of interactions and interviews were combined into episodes, and initially classified on the basis of the mathematical audit. Each episode comprised a description of the activity, the strategies used and any substantive issues that arose. The majority of episodes were unproblematic and routine to practitioners: indeed, in some cases, the strategies used were almost subconscious, and subjects found it hard to describe them, at the time or when probed after the event. However on some occasions we were able to gain insight into the approaches used in routine activity through spontaneous exchanges between subjects and their colleagues.

In contrast, we observed a number of 'non-routine' episodes that, from an early stage, appeared radically different in that they involved 'breakdowns' indicated by conflict, disagreement or doubt. Although rare, these episodes provided an illuminating view of practitioners' thinking about particular problematic situations and the knowledge mobilised to resolve them.

1.3. Simulation interviews

A series of task simulation interviews was devised, based on a selection of workplace activities and contexts. The majority of the simulations recreated the breakdown episodes witnessed during the ethnographic observation. Each scenario was made as 'real' as possible, by the use of context-rich descriptions of the activity, with supplementary information provided in the form of familiar resources and visual displays; e.g. copies of blood pressure charts, screen shots of instrumentation. The interviews were developed with the support of senior staff to ensure their *prima-facie* validity. In the nursing study, three scenarios were recreated (infusion flushing, interpreting time-series data and drug timing); in the flying study, two scenarios were used (crosswind landing and in-flight heading adjustment).

These interviews provided a sensitive instrument with which to explore the reasoning of subjects in a way more detailed than was possible during the *in situ* observations and discussions. They also allowed us to explore connections between context and knowledge by manipulating the relationships in the episode in ways that moved the simulation away from familiar ground. Thus, responses to initial questions on each simulation related directly to the relevant episode (what would the subject do and why?), providing information on how any relevant mathematical elements were used. Subsequent questions were increasingly novel and generalised, involving situations our subjects were unlikely to have encountered at work. This process of progressive mathematisation shifted from the familiar to the unfamiliar in several ways: inverting knowns and unknowns, using unfamiliar representations, or setting familiar problems in unfamiliar contexts while retaining elements of the same mathematical structure.

1.4. Questionnaires and teaching experiments

In the nursing study, we developed questionnaires based on two of the scenarios in the simulation interviews to investigate our findings with a greater number of subjects. In the banking study, questionnaire items were based on activities identified in Stage 1. Both questionnaires were administered to all the practitioners recruited for the teaching experiments.

We distinguished some mathematical elements in a selection of the key activities we had identified in the ethnography, and used these as a basis for teaching experiments with both the nurses and bankers. We had two aims: first, to give practitioners a set of tools with which to reflect upon and construct mathematical models of some workplace scenarios; and second, to provide us with a window through which to view their thinking. We were given the opportunity to carry out the teaching experiments as part of existing training programmes. In the banking study, we developed a three-day course based on modelling a range of financial instruments through programming, using a dialect of Logo called *Microworlds*. In the nursing study, we planned a teaching experiment around two modelling activities: the analysis of cross-sectional vital sign data and its relationship to age and sex, and an exponential model of drug dose levels in the blood. Both sets of activities involved the use of the data analysis software *TableTop*.[3]

1.5. Data Collection and Analysis

The focus of our research was on the mathematics used by confident and relatively experienced practitioners, and not the mathematics used by novices to the professions. We therefore recruited as many subjects as possible in each group, where all were volunteers who had worked in the profession for at least two years. Table 1 below summarises the range and type of data collected.

Methods		Nursing Study	Banking Study	Flying Study
Documentary Analysis	*No. Texts*	6	6	3
Interviews with Senior Staff	*No. Senior Staff*	4	2	1
	Total Hours	8 hrs	4 hrs	4 hrs
pre-Interviews with Subjects	*No. Subjects*	12	6	—
	Total Hours	12 hrs	6 hrs	
Ethnographic Observation	*No. Subjects*	12	6	5
	Total	80 hrs	16 hrs	12 hrs
Hours				
Simulation Interviews	*No. Subjects*	5	—	5
	Total Hours	10 hrs		7.5 hrs
Questionnaires	*No. Subjects*	28	20	—
Teaching Experiments	*No. Subjects*	28	20	—
	Length	6 hrs-1 day	18 hrs-3 days	
	No. Sessions Run	2	8	

Table 1. Summary of data collected for each study

In this chapter, we draw on evidence mainly from the nursing and pilot studies to make some general points about how mathematics is used at work and what this might indicate for useful paradigms of research in this area[4].

2. VISIBLE MATHEMATICS AND ROUTINE ACTIVITY

To begin our analysis, we require a definition of what we termed the 'visible mathematics' of the workplace, identified in our documentary analyses and interviews with experts. The visible mathematics of a practice is derived from school mathematics. It includes conventional mathematical symbolism and representations (e.g. number, graphs, tables, scales, algebraic formulae, and geometric drawings), but also the use of concepts, strategies and methods of the mathematics classroom. It should be stressed that visible mathematics means visible *to us*, and one of our first questions was to examine how far practitioners have similar perceptions. The notion of visible mathematics allowed us to construct a first but necessarily incomplete map of the mathematics 'located' in each occupation in terms of a well-defined set of mathematical elements and strategies.

The documentary analysis revealed some interesting differences in the presentation of the visible mathematics of our three practices. For example, the mathematics of finance textbooks tended to be presented with little indication of the context in which mathematics might be used, while textbooks in aviation and nursing are more oriented to the implementation of particular activities — although some nursing texts also provide materials for developing the 'pre-requisite mathematics' in areas such as basic arithmetic. Examples of items in each audit are given in Table 2.

Nursing Study	Banking Study	Flying Study
DRUG PREPARATION single and repeated doses; infusions; finding doses from prescription and drug concentrations; dilution issues; measurement; ratio and proportion.	PRESENT AND FUTURE VALUING simple and compound interest; finding future and present values; percentages; inverting.	HEADING ADJUSTMENTS setting heading to take account of wind velocity; ground and true air speed; triangle of velocities.
INFUSION MONITORING coordinating infusion rate, time and volume; checking concentrations; measurement; ratio and proportion.	COMPOUNDING AND FREQUENCY compounding over different frequencies (e.g. annual, quarterly); non-linear functions and graphs.	WIND LANDING LIMITS separating components of wind velocity; checking cross and headwind limits.
FLUID INTAKE AND OUTPUT measuring hourly fluid intake and output; recording and updating fluid balance charts.	BOND PRICING present value pricing of multiple-payment bonds; geometric sequences.	PRE-FLIGHT FUEL CHECKS optimising fuel load, taking account multiple factors; e.g. weather, wind speeds, fuel prices, passenger loads, landing and take-off load limits.
VITAL SIGNS & LAB DATA measurement of vital signs (blood pressure, etc.); recording on time-series graphs; scales; interpreting lab report data; index form.	PRICE RISK risk due to changes in interest rates; difference functions; rates of change; differentiation.	IN-FLIGHT FUEL CHECKS measuring, updating and monitoring fuel burn rates.

Table 2. Mathematical audit from texts and senior staff interviews

Examination of the content and strategies of visible mathematics reveals that it was largely concentrated in two types of activity: either finding solutions to particular problems in procedural ways through the use of specialised algorithms (e.g. doses using a drug formula, 'present values' with built-in spreadsheet functions, aircraft heading and speed with a navigation computer[5]); or carrying out routine data gathering and representation by measuring, recording and calculating particular quantities (e.g. measuring and plotting hourly vital sign data, calculating and recording hourly fuel-burn rates).

From the ethnographies, we observed that a substantial proportion of routine episodes involved a variety of simulations of functional relationships of one or more variables to give a direct way to calculate a required unknown. Frequently practitioners used personal, mental approaches or 'look-up' methods rather than the standard procedures of visible mathematics. Mental methods were used either to find exact answers to particular problems (e.g. mental calculation of drug doses), or to estimate a solution on the basis of a simplified model of the situation (e.g. calculation of maximum drift in flying using only some of the available data, or estimating a normal blood pressure from the age of a patient). Look-up methods involved using different types of tables and charts to derive an unknown quantity from one or more known quantities (e.g. nomograms to find body surface area from weight and height).

To concretise the discussion, we take an example from the flying study. First we must note that any ethnography involves learning the language of the workplace and the meanings of unfamiliar, even familiar words. Pilots frequently have to calculate the crosswind — the component of wind that acts at right angles to the direction of travel and blows an aeroplane off course. We give an example of an episode where this measure was needed in order to decide whether it was safe to land the plane in wintry conditions: that is, the pilots had to use wind velocity information to decide whether the crosswind on the runway exceeded set limits. Flight procedures dictate that if wind velocities exceed these limits, the pilot should fly to an alternative airport. The limits are given in terms of crosswinds and head/tailwinds, with different limits depending on the 'breaking actions' of the runway; i.e. whether the runway is icy, snowed over, gritted, etc. (see Table 3). For example, if the friction on a runway is 0.33, it has a 'medium' breaking action and hence has a crosswind limit of 15 knots and a tailwind[6] limit of 5 knots. The situation can be more complex with long runways, whereby different breaking actions are given for different sections. In such cases, the standard procedure is to take the worse case; e.g. if three sections of a runway are designated 'Poor-Medium-Medium', the whole runway is assumed to be 'Poor'. Pilots therefore need to use the direction of the runway and their latest wind velocity on the ground to calculate tailwind and crosswind, and check whether or not these exceed the limits.

Coefficient of Friction	Breaking Action	Tailwind	Crosswind Limit
0.25 or below	Poor	0	5
0.26 to 0.29	Medium-Poor	0	5
0.30 to 0.35	Medium	5	15
0.36 to 0.39	Medium-Good	5	15
0.40 and above	Good	10	33

Table 3. Breaking actions for the Fokker 50 aircraft

In this episode, we observed a pilot and co-pilot descending towards runway 05 (which means it had a bearing of 050°). The latest wind velocity was 13 knots with a bearing of 100°. Although the wind was not strong, the breaking action on the runway was 'Poor', giving a crosswind limit of 5 knots. The captain needed to check whether the wind velocity exceeded the crosswind limit. First, he calculated the angle off-the-nose; i.e. wind direction relative to the direction of the runway. This was simple for him and he knew immediately it was 50° in this case (see Figure 2). We note that this calculation is by no means easy for non-pilots. He then had to estimate the value of the crosswind.

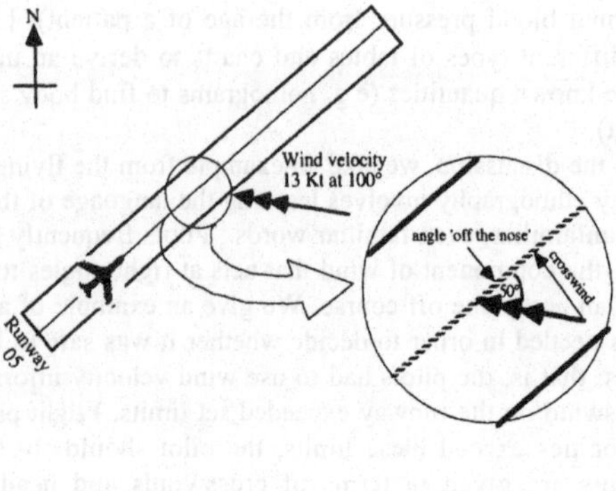

Figure 2. Schematic diagram of the surface wind on runway 05

The captain used a quadrant diagram (see Figure 3), specifically designed to find the runway crosswind. He described his procedure as follows:

> Pilot: We've got to go and make sure we haven't got a cross-wind limit over that amount [pointing to 5 on table in row next to poor breaking action]. So that's 50 degrees off at 13 knots. Just roughly, that's giving a crosswind of about 10 knots.

He found his estimate of 10 by drawing round 50 degrees from 13 on the vertical axis and reading off 10 on the horizontal axis (as shown in Figure 3). Now a crosswind of 10 was over the limit, so the pilot and co-pilot discussed what to do. In the end, they decided to land the plane anyway, justifying this by saying:

> Pilot: …They always take the worse case. Maybe we have to start thinking about taking another runway or going somewhere else – but it looks OK.

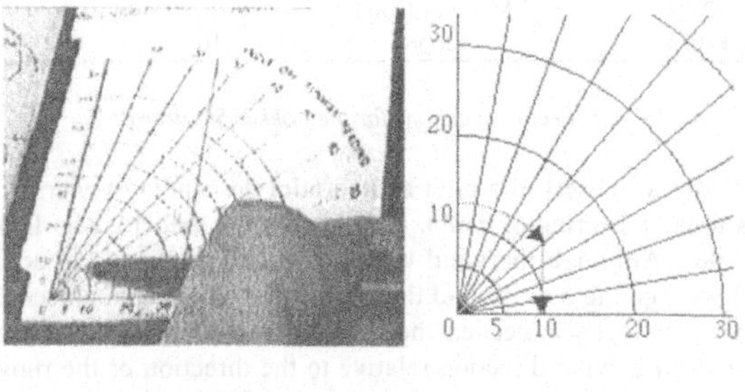

Figure 3. Screen shot and schematic detail of pilot estimating landing crosswind using a quadrant diagram

In this episode, the look-up method used clearly circumvented the need for working directly with any trigonometric function. However, the pilot had to work out the angle off-the-nose and coordinate this value with information on wind velocity and breaking action to come to a decision.

Another intriguing pilot's method for estimating crosswinds involves using a watchface to convert between the angle off-the-nose and the proportion of the wind speed that makes up the crosswind component. For example, if the wind is 20 knots and the angle of the wind off-the-nose is 45°, this is interpreted as follows: 'It is the 45th minute which is three quarters around the watchface, so the crosswind component is approximately three quarters of the wind speed, namely 15 knots'. The mathematical basis for this method is a linear interpolation of the sine function between 0° and 60°, which ranges between 0 and 1. Clearly, the method is accurate for angles 0° and 30°, but becomes less accurate as the angle approaches 60° (see Figure 4). Pilots 'know' this, so above this angle they tend to interpret the wind velocity as almost all crosswind; which again is a very reasonable estimate in practice.

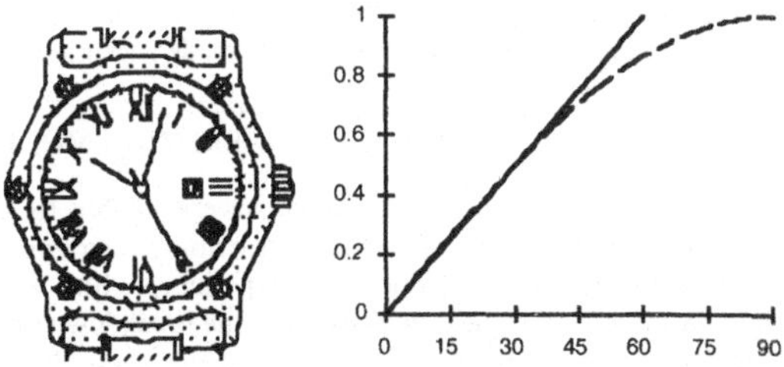

Figure 4. Watchface and linear interpolation of sine function

Pilots repeatedly encounter the crosswind problem, so they have to develop at least one method that generates a solution within an appropriate degree of accuracy. It is also an integral part of their initial training when they are introduced to trigonometric methods. But what became clear from our study was that even those pilots who were very confident with trigonometry never used it. Rather they relied on mental or look-up methods that they had picked up 'on the job'. Even in interviews away from the cockpit, none of the pilots used trigonometric functions when working with this type of problem or any other problems where, from a mathematical perspective these would provide the obvious methods.

Why? Clearly, making an approach for landing is an intensive activity that involves all of a pilot's cognitive, perceptual and sensory-motor faculties. As they make the approach and move closer to the runway, they need to be in a state of readiness to abort the landing if they think it is dangerous. In unstable weather conditions where they are being continuously bombarded with the latest runway wind velocities and breaking actions, they need to be able to make a judgement relatively quickly - to land or to pull up. During navigation in the sky, decisions can be re-evaluated based on instrument feedback and appropriate action taken to re-adjust. Landing gives relatively little room to re-evaluate a solution. Thus, rules of thumb which quickly and efficiently answer the question 'to land or not' are seen as more effective than general-purpose procedures involving trigonometry – especially when these latter calculations provide unnecessary accuracy.

So pilots use methods rather different from those of mathematics. In the ethnography and flying study interviews it became increasingly clear that pilots rarely inhabit the mental space anticipated by visible mathematics, i.e. of wind and plane vectors constructed in triangles of velocities. Rather, they exploit a different set of objects connected in different ways. Pilots imagine themselves in the plane and use *relative* wind direction ('angle off-the-nose'), crosswind and tailwind components, while absolute quantities, such as heading, are invisible. Intuitive estimates of the measures they need are supplemented by disparate rules of thumb acquired in the practice which work in limited but well-defined sets of circumstances (such as the watch rule). For calculations to be meaningful, we conjecture that the 'feel' of being in an (even imaginary) cockpit is crucial, as it is from this position that judgements and estimations are made successfully.

3. BREAKDOWN EPISODES AND UNDERLYING MODELS

In some cases during our ethnographies, subjects were unable to articulate any strategy when asked how they solved a problem, describing their approach in terms of 'having a feel for the answer', 'intuition' or 'experience'. But, we were able to throw some light on these approaches by analysis of the breakdown episodes, where activity took on a character altogether different from the unproblematic nature of routine episodes. Breakdowns apparently involved very little visible mathematics and decision-making seemed at first sight to be based only on professional judgement or recourse to concepts or judgements outside the mathematical domain. However, further investigation revealed that intertwined with these judgements were mathematical elements — *but not necessarily those of visible mathematics*. These elements often comprised some unpredicted factors that underpinned their 'intuitive' approaches to routine practice or fitted a model that was hidden when there was no need for its articulation. Conflict provoked practitioners to justify their approaches and the models on which they were based, although it must be said that we have no evidence that practitioners were aware of the models underlying their descriptions. We illustrate the point in relation to a breakdown episode observed in the nursing study.

The interpretation of vital signs, such as blood pressure, involves knowledge of a patient's clinical background, age, condition and recent treatment, all of which relate in a complex way to deciding whether a patient's condition is 'normal'. One simplified model of the normality of a patient's blood pressure is based on how this relates to a population, which requires both an appropriate definition of the target population and at least some elements of the statistical machinery of frequency distributions such as mean, standard deviation, type of distribution and so on. From this population model, judgements can be made regarding how a particular patient's blood pressure deviates from the mean blood pressure for the appropriate target population, where 'abnormality' can be partly defined on the basis of standard deviation cut-off points either side of the population mean.

This is a highly simplified model of which defining an appropriate population is a crucial feature: for example, blood pressure increases with age, so any model applied to a child's blood pressure needs at least to define the population as within the child's age group.[7] More accurate models can be defined by taking further patient characteristics into account and defining the population more stringently. However, medical statistics are only available at a relatively crude level of discrimination, such as in terms of age and sex. This, together with the fact that blood pressure varies due to many transitory 'non-critical' factors such as emotional state, physical activity, level of hydration, etc., quickly makes it clear that population models can only serve as useful guidelines and that other kinds of knowledge need to come into play for appropriate judgements of normality to be made. In our ethnography on the wards, we came to view the question of what this knowledge was and how it related to population models as increasingly important, as it provides an example central to nursing practice of how mathematised descriptions of reality might become interpreted and acted upon.

It was around this issue that we observed a breakdown episode based on a difference of opinion over the 'normality' of a child's blood pressure. The child was on a drug that could not be given if the blood pressure was too high. Thus the issue of whether or not the drug should be administered could offer some insights into the way 'normality' might become a highly disputed territory and how mathematical models played a part in decision making. The initial part of the episode involved a nurse taking regular readings of the patient's blood pressure, many of which she considered abnormally high. Her initial readings at 3:50 pm and 4:15 pm were 131 over 82 and 128 over 80 respectively, using different equipment to check whether this was due to instrumentation error. Half an hour later, the blood pressure had gone down to 123 over 73 - a reading she still considered alarmingly high. Unsure about whether to give the drug, she finally consulted two doctors - the registrar and the senior house officer (SHO):

> Nurse: Well, I've just done it, and I'm about to do it again. 123 on 73. So, do I give it, do I not give it?
>
> Registrar: How old is he?
>
> Nurse: He is, mmm..., 2 at the end of this month.

The doctor clearly agreed with the nurse that the blood pressure was abnormally high, and asking for the age was an indication that she wanted to make sense of this reading with respect to the population. The doctor then decided to change the prescription.

Registrar: Some people think this isn't the best treatment if it's due to his [inaudible] anyway.

Nurse: No? Let me get his chart then, in case you want to change it.

Registrar: (to Senior House Officer) I'm interfering with your prescription.

SHO: That's alright, feel free.

The registrar finally attempted to reassure the nurse by placing the child's blood pressure within a half-remembered percentile for children of similar age.

Registrar: Let's watch and wait and see if it all...

Nurse: Yeah?

Registrar: Well, I think.... the 97th percentile for normal blood pressure at the age of 15 months is ..Is something like 118 on 70 because I looked it up once. So he's actually a bit older than that, so I'm sure ... [inaudible] so let's watch it and see. Review it on a daily basis.

Nurse: OK.

Clearly, the doctor was using a population model as a way of making sense of the child's blood pressure, as least in her communication to the nurse. And, if the registrar was right, the blood pressure was very high with respect to the population of toddlers. What is less clear is whether the nurse used a similar model when she made her initial evaluation, especially as we had never witnessed the language of centiles and other fine-grained descriptors of distributions in any other part of the study. Where population norms were described by nurses, they used the more simplified representation of upper and lower bounds with different ranges associated with children of different ages.

This breakdown indicated that the nurses used another concept of 'normal' to inform their decision. This we termed an individual's 'baseline blood pressure' that does not necessarily fit easily into a population model. Baseline blood pressure involves associating a 'normal' blood pressure reading with a particular patient after sustained contact over time. It is measured when the patient is regarded as stable and in a relaxed state. It is this statistic that nurses use to judge whether a patient has deviated from normality, and it is only combined with population summary statistics in a fragmentary way when this is seen 'to help' the nurse's argument. Having uncovered this nurse's view of normality in the breakdown episode above we were able to identify further instances of the same phenomenon in interviews and observations. For example, in the following interview extract, a nurse was asked to examine the blood pressure of a fictitious patient over time. Immediately, she referred to the individual's baseline pressure:

Int: What's the greatest increase in the systolic pressure over this 24 hours?

Nurse: Well I would probably have a look at what her baseline was. [...] So you know that, say, her normal would be, just for arguments sake say, 110 over 55 so you know that that's her normal, to say that's what it was on admission. So you take that as a baseline and then you look at that and you think 'well that's 120 on 60' and you think 'oh yeah' she has had an increase in blood pressure.

This illustrates further how the dominant underlying models for nurses are based on the notion that each patient has a relatively unique physiological profile, and that decision-making (when making judgements about abnormality) should, at least in part, take account of this profile. By establishing a patient's normal profile of vital signs, nurses are in a better position to judge when a patient has undergone significant changes in their condition. Nurses, of course, do not treat populations but individuals. While from a mathematical point of view, there is no difference between the notion of a mean derived from a population (e.g. 'what is the average blood pressure of 30 year-old males?') and an individual (e.g. 'what is the average blood pressure of *this* 30 year-old male?'), there is a considerable difference in terms of the specific objects and relationships under scrutiny.

We gained further insight into this phenomenon in the nurses' teaching experiment, where they explored vital sign data. Predominantly, their explorations focused on the unexplained variation rather than the correlations with background factors, such as age and gender. It was this variation rather than the trends or patterns that was the focus for nurses' 'explanations'. We would argue that this orientation is at least partly traceable to nursing practice, where variation and its explanation is often the catalyst for action. If variation cannot be explained, the nurse needs to call a doctor to discuss treatment. (For further details on nurses' meanings of average and variation, see Noss, Pozzi and Hoyles, 1999).

4. MATHEMATISING WORKPLACE ACTIVITY

Our studies showed that the visible mathematics of practice was almost invariably associated with routine activities, at least for the experienced staff that we observed and interviewed. These activities often involved measurement and recording, or the use of algorithms to find unknowns from one or more known quantities. Most of these methods coincided with those described by senior staff or in the teaching texts, and would not have been out of place if found in a mathematics classroom. However, there were a significant number of methods even in routine activity where practitioners used a range of apparently idiosyncratic mental strategies finely-tuned for solving particular problems in very specific circumstances. Despite their specificity, we could discern sound mathematical models underpinning these strategies. Many of these strategies only came to light during workplace observations and would not be mentioned by practitioners as 'using mathematics at work', suggesting that practitioners were either unaware that they used them, or did not regard them as sufficiently 'mathematical' to be of interest to us. A second group of methods we observed were the look-up strategies or use of tools (for example the crosswind method described above, for evidence from the nursing studies on their use of resources see, Pozzi, Noss & Hoyles, 1998) which essentially circumvented the need for general mathematical solutions to a problem.

Looking more generally, the associated mathematics of routine activity consisted of an assortment of methods and algorithms to solve particular, well-understood problems. Whether methods were drawn from the institutionalised mathematics of school or the personalised mathematics of the occupational groups, the common feature would seem to be the scope to solve problems quickly and efficiently. Sometimes this may involve visible mathematics, but it could easily utilise one of the many 'tricks of the trade'; the finely tuned mental strategies or various tailored look-up methods that develop within any working culture.

From a mathematical point of view, efficiency is usually associated with a general method that can then be flexibly applied to a wide variety of problems. This is clearly not the case in the workplace. Even if a number of tasks could potentially be solved with a similar approach, practitioners prefer to use different approaches for each task, partly based on the resources at hand. The crucial point is that orientations such as generalisability and abstraction away from the workplace are not part of the mathematics with which practitioners work. Thus, mathematical routines are rarely, if ever, interpreted as exemplars of completely general mathematical concepts or relationships, nor are they manipulated or transformed to solve different types of problem. By the same token, the tables and charts of lookup methods encode appropriate quantitative or geometric relationships, but render them unavailable to scrutiny.

For the practitioners in our studies, the computational and estimation methods of routine activity - in all their many forms - were more than adequate for their purposes.[8] Also, most subjects could answer familiar, routine questions correctly even when removed from the workplace. Thus, in the nursing and pilot interviews, subjects made confident use of written arithmetic algorithms, mental approaches, look-up methods and the use of specialised and generic tools (e.g. the navigation computer, wind quadrant, basic calculator). Yet breakdowns in workplace routine revealed another face of mathematics intertwined with professional expertise, with judgements frequently based on mathematisable models of the situation that we had not predicted and which included both qualitative and quantitative elements. They may not be general away from the workplace but they were certainly not simple routines or reactions to immediate concerns. For example, the notion of an individual baseline described in the nursing study is amenable to mathematical description: for the nurses however, the concept seemed only available at the level of professional knowledge.

We have reported other examples where we could discern some kind of mathematical model underpinning practice, but one with distinctive workplace features. The model comprises an abstraction from the immediacy of the situation, but because of these workplace features it retains elements of the setting — hence we have called practitioner's conceptions of the mathematics they use at work, *situated abstractions*. For example, some nurses' concepts of overdosing involve discrete notions of daily recommended doses which prove unhelpful when giving drugs at irregular frequencies (see Hoyles, Noss and Pozzi, 1999). When calculating drug doses nurses call upon a range of proportional strategies that, though correct, frequently depend on the drug in question (see Hoyles, Pozzi and Noss, in press). In the banking study, time is conceived 'backwards' as time to investment maturity, and a subject's notion of time is contingent on arbitrary historical aspects of different financial instruments, where even the number of days in a year is not fixed (see Noss and Hoyles, 1996b). All this evidence from our simulation interviews leads us to the conclusion that practitioners' epistemologies of quantity, space and time can differ fundamentally from those anticipated from a school mathematics orientation. Practitioners' epistemologies are based on intimate connections with the specificities of their practice. Moreover, these models often remain at an intuitive level, hidden until a breakdown occurs when conflict provokes practitioners to justify their strategies.

So, how flexible are these workplace strategies? Our evidence suggests that workplace strategies do often require some degree of personal identification with the particularities of the setting — an *anchor* — for the coordination of mathematical and non-mathematical elements. For example, we have seen how nurses' calculations of average do not exist as entities separate from the qualitative features of the patients themselves. More revealing still, when we faced pilots with estimating cross current values in a problem involving a boat rather than a plane, few had any useful intuitions about the solution and were unable to set up their crosswind and tailwind components in view of the different scale of the quantities involved.

Within a familiar setting, fragments of knowledge seemed to be coordinated in work by anchoring meanings in highly specific facets of the practice. Provided this anchoring is available, the setting can be re-created and a synergy constructed between the mathematical and professional knowledge — even in non-workplace settings. When this anchoring is not available, responses to unfamiliar questions tend to involve attempts to apply a fragment of professional knowledge, a half-remembered rule from school mathematics or a novel, though generally unsuccessful, use of a familiar tool. In teaching experiments, practitioners were able to re-express routine practices through the use of mathematical models and in some cases did so successfully. However, this was not straightforward, as there was frequently a mismatch between practitioner models of practices and those constructed from a mathematical orientation.

The message from our research is that practitioners *do* use mathematics in their work, but what they use and how they use it may not be predictable from considerations of general mathematical methods. Moreover, strategies depend on the nature of the activity, whether it is routine or breakdown, and the resources available.

Specifically, we suggest that there are three ways to classify the relationship between mathematics in use and mathematics *per se*. First, practitioners may deal with the same objects and relations, but in ways and with a workplace language, which differ from that of established mathematics. Second, the objects and relationships distinguished by practitioners and the models they conceptualise for their activity may not match those of standard, visible mathematics: that is, while their conceptions achieve generality within the work setting, they may not be the same as those of mathematical discourse (e.g. cross/tail wind rather than vector triangles). Third, the generality of workplace mathematics may be limited to specific and well-defined sets of circumstances that work in situ but nowhere else. Clearly, if we are to successfully mathematise workplace activity we need to be aware of all three possibilities, and seek to identify and analyse the situated abstractions developed by practitioners in their workplace. Simply looking for visible mathematics is inadequate. But so too, we suggest, are some studies in the tradition of situated cognition, which fail to discern the breadth and richness of the mathematical models in use.

5. REFERENCES

Barrallo, J. (1998). *Mathematics and Design 98*. University of the Basque Country.

Fahrmeier, E. (1984). Taking Inventory: Counting as Problem Solving. *Quarterly Newsletter of the Laboratory of Comparative Human Cognition* (6), 6-10.

Fitzgerald, A., Purdy, D.W. and Rich, K. M. (1981). *Mathematics in Employment (16-18)*. Bath.

Gagliardi, P. (1990). *Symbols and Artifacts: Views of the Corporation Landscape*. New York: Aldine de Gruyter.

Hall, R., and Stevens, R. (1996). *Teaching/ Learning Events in the Workplace: A Comparative Analysis of their Organisational and Interactional Structure*. Paper presented at the 18th Annual Conference of the Cognitive Science Society.

Hall, R., and Stevens, R. (1995). Making Spaces: a Comparison of Mathematical Work in School and Professional Design Practices. In S. L. Star (Ed.), *The Cultures of Computing* (pp. 118-143). London: Basil Blackwell.

Hancock, S. J. C. (1996). *The Mathematics and Mathematical Thinking of Seamstresses*. Paper presented at the AERA National Convention, New York.

Harris, M. (1987). An Example of Traditional Women's Work as a Mathematics Resource. *For the Learning of Mathematics, 73*(6), 26-29.

Harris, M. (1991). *Schools, Mathematics and Work*. Brighton: Falmer Press.

Hoyles, C., Noss, R., and Pozzi, S. (1999). Mathematising in Practise. In C. Hoyles, C. Morgan and G. Woodhouse, (Eds.), *Rethinking the Mathematics Curriculum*. London: Falmer Press.

Hoyles, C., Pozzi, S., and Noss, R. (in press). Proportional Reasoning in the Workplace: A Study of Expert Nursing Practice. *Journal of Research in Mathematics Education.*

Lave, J. (1988). *Cognition in Practice: Mind, Mathematics and Culture in Everyday Life*. Cambridge: Cambridge University Press.

Lave, J., and Wenger, E. (1991). *Situated Learning: Legitimate Peripheral Participation*. Cambridge: Cambridge University Press.

Leont'ev, A. N. (1978). *Activity, Consciousness and Personality*. Englewood Cliffs, NJ: Prentice-Hall.

Masingila, J. O. (1994). Mathematics Practice in Carpet Laying. *Anthropology and Education Quarterly 25*, 430-462.

Millroy, W. L. (1992). An Ethnographic Study of the Mathematical Ideas of a Group of Carpenters. *Journal for Research in Mathematics Education, Monograph 5*.

Noss, R., and Hoyles, C. (1998). New Numeracies for a Technological Culture. *For the Learning of Mathematics, 18*(2), 2-12.

Noss, R., and Hoyles, C. (1996a). *Windows on Mathematical Meanings: Learning Cultures and Computers*. Dordrecht: Kluwer.

Noss, R., and Hoyles, C. (1996b). The Visibility of Meanings: Modelling the Mathematics of Banking. *International Journal of Computers for Mathematical Learning, 1*(1), 3-31.

Noss, R., Pozzi, S., and Hoyles, C. (1999). Touching Epistemologies: Meanings of Average and Variation in Nursing Practice. *Educational Studies in Mathematics, 40*, 25-51.

Nunes, T., Schliemann, A., and Carraher, D. (1992). Ethnomathematics and Everyday Cognition. In D. A. Grouws (Ed.), *Handbook of Research on Mathematics Teaching and Learning* (pp. 557-574). New York: Macmillan.

Nunes, T., Schliemann, A., and Carraher, D. (1993). *Street Mathematics and School Mathematics*. Cambridge: Cambridge University Press.

Pozzi, S., Noss, R., and Hoyles, C. (1998). Tools in Practice, Mathematics in Use. *Educational Studies in Mathematics, 36*, 105-122.

SCAA. (1997). *Literacy and Numeracy in the Workplace.* York: School Curriculum and Assessment Council.

Scribner, S. (1984). Studying Working Intelligence. In B. Rogoff and J. Lave, (Ed.), *Everyday Cognition: Its Development in Social Context* (pp. 9-40). Cambridge MA: Harvard University Press.

Scribner, S. (1986). Thinking in Action: Some Characteristics of Practical Thought. In R. J. Sternberg and R.K. Wagner, (Eds.), *Practical Intelligence: Nature and Origins of Competence in the Everyday World* (pp. 13-30). Cambridge MA: Harvard University Press.

Smith, J. and Douglas, L. (1997). *Surveying the Mathematical Demands of Manufacturing Work: Lessons for Educators from the Automotive Industry*. Paper presented at the AERA Annual Meeting, Chicago, IL.

Wolf, A. (1984). *Practical Mathematics at Work: Learning Through YTS (Research and Development Report No. 21)*. Sheffield: Manpower Services Commission.

6. NOTES

[1] We acknowledge the support of the Economics and Social Science Research Council (ESRC) in funding the project 'Towards a Mathematical Orientation through Computational Modelling' (Grant No. RO22250004).

[2] Mathematical texts were available for all three groups. Nurses and pilots have well-established initial training programmes which include the assessment of mathematics. Thus a number of self-study and teaching texts are available in these areas. Investment banking has diverse access routes and more differentiated mathematical requirements. However, there is still a wide range of texts available on finance mathematics at many levels.

[3] Tabletop was developed at TERC (Cambridge MA) and is distributed by Broderbund Software, Novato, CA USA.

[4] Our conclusions will draw on all three studies: more discussion of the banking study can be found in Noss & Hoyles, 1996b).

[5] Navigation computers are not in fact electronic as the name suggests, but manual hand-held devices made up of a slide rule and transparent rotating discs. They can be used in a number of ways for marking wind velocities, aeroplane headings and speeds.

[6] Given that aircraft usually land into the wind, there is effectively no tailwind, so the crosswind component is normally the only concern.

[7] This is akin to judging whether a child's height is 'normal' - such a judgement would always be interpreted as 'normal for their age-group'. Another example is judging whether an adult's weight is normal, which usually takes account of height.

[8] Some subjects did however describe how they needed to 'refresh' their mathematics at various stages of their careers, particularly when they needed to pass various occupational or recruitment assessment tests; many of which involved some mathematical items. Interestingly, subjects often expressed a strong sense of incredulity and scepticism with respect to the validity of such tests.

CHAPTER 3

FORMS OF MATHEMATICAL KNOWLEDGE RELATING TO MEASUREMENT IN VOCATIONAL TRAINING FOR THE BUILDING INDUSTRY

MADELEINE EBERHARD[1]

Abstract. It is of crucial importance when setting up work on the construction site to respect the measurements laid down. A specific measuring technique is used for this purpose, namely cumulated dimensions. In the context of vocational training, the effectiveness of this technique may be demonstrated mathematically and may thereby provide a useful contribution to learning about the conditions under which it is used.

Analysis of observations carried out in a French technical high school reveals the constraints which make such use of mathematics improbable at the present time. Teachers of 'construction' show in action the economies to be made through this technique, laying stress on its efficiency. It is not their responsibility to teach the elementary mathematics which prove that efficiency. In mathematics teaching the basic concepts needed to justify the technique are indeed present, but contained within mathematical structures which have no relation to problems involving errors of measurement.

The purpose of this article is to study the relationship between the aims of institutions which carry out vocational training for trades in the building industry and the nature of the mathematical knowledge taught. What is the nature and status of the mathematical knowledge which comes into play, even implicitly, during the course of professional activity on the building site? Within training establishments is such knowledge considered to be useful? Is it taught explicitly? If so, by whom and how? This article proposes to explore these questions in the context of tasks for which *foremen* or *skilled workers* may have responsibility.

1. COMBINING TWO APPROACHES TO KNOWLEDGE: THE CONSTRUCTION SCHOOL

This study takes the form of an investigation centring on a training course for the building construction industry (the construction school) which was attached to a technical high school.[2] Direct observation on the building site forms the main element of the study. A contract had been placed to build garages for a housing development which was under construction, and the job of the construction school was to complete the task, in a way that was acceptable to both the architect and the

Bessot & Ridgway (eds.), Education for Mathematics in the Workplace, 37—51.

client. It was a setting in which a professional project was being carried out within the bounds of the usual contractual limitations; it was a place of learning for pupils in their final year of a Technical High School preparing for the Vocational Diploma (Brevet de Technicien) of Building Site Supervisor[3].

Corresponding to the dual aims of the construction school, knowledge has a dual function within the institution: it is treated differently according to whether it relates to solving practical problems or to teaching.

As a result, in order to analyse the function of knowledge in the construction school, we may make a distinction between knowledge required when working on practical problems and that which fulfils a theoretical function[4]:

> When it fulfils a theoretical function, knowledge serves to produce knowledge; and when it has a practical function, knowledge serves to produce results in the domain where it is a tool for action. (Mercier, 1995, p.130).

In order to analyse the status of knowledge in this teaching institution (as a Vocational Diploma (BT) class within a Technical High School), we shall use ideas from Chevallard (1991). Within the theoretical framework of didactical transposition Chevallard draws a distinction between types of knowledge according to whether or not they are in the teacher's field of *didactic perception*. Knowledge in didactic perception corresponds to:

-topics taught which are considered by both teacher and pupil, in the context of the didactic contract[5], to be required knowledge.

-topics learnt which are not directly assessed by the teacher, but which may occasionally be subject to judgement on whether or not they have been understood.

2. THE CENTRAL ROLE OF MEASUREMENTS IN PRACTICE

According to one of the teachers of building construction:

> at the end of the final examination, BT pupils will leave as skilled workers or as foremen, since they will have done many 'workshops' [these fall within technological education of which the construction school forms a part]. They will be able to enter construction site employment immediately.

At the end of their training, these pupils are therefore considered to have acquired sufficient knowledge of basic building site practice. A good many of these basic practices are to do with 'Setting out'. This surveying term refers to all the activities concerned with marking out appropriate positions on the building site, on the basis of the architect's plans[6].

> This consists of marking out, in succession, on site: a) the boundaries of the property; b) the contours of the general excavation where building is to take place; c) the lines and contours of the actual excavations and of the foundations.[...] It also consists of positioning the various components or parts of the building according to height level (Olivier[7], 1976, p 19).

Setting out determines the lines or contours of the foundations to be constructed, as well as the location of future walls and columns. It is done by making marks on the ground and by locating setting out pegs. A setting out peg is a piece of wood with a nail in the top, aligned vertically and held in place with two stakes; the pegs are joined with string to show building lines. Other sorts of markers, such as a line of paving slabs, might also be used. These marks must express, within the space set aside for construction, the measurements laid down in the architect's plans. The *transfer of the dimensions* will indicate for practical purposes the results of this marking out. Accurate setting out is essential for the successful completion of building work. It requires a mapping from a technical drawing to the site; mathematical skills in interpreting the dimensions required; and practical skills in locating, measuring, and checking the work.

> In fact, the basic operation for the technician responsible [for setting out] consists of *marking out alignments* [...]. All these operations [of setting out] [...] necessarily entail the exact transfer: - of distances; - and of angles, in order to obtain a precise representation of the building to be constructed. (Olivier, 1976, p.19)

This notion of ‘exact transfer’ defines an ideal[8] of working in order to fulfil obligations contracted between the architect, client and building company. Measurements play a central role in determining the success of a project; they are also of prime importance when judging the *quality* of the finished product.

Exact transfer can never be attained and the inevitable *imprecision* of actual measurements is countered by the notion of *tolerance*. Tolerance limits uncertainty by specifying the various error allowances for measurements. The need for accurate measurements, within the limits of tolerance, is a principle of all work on the building site; but it is absolutely essential in setting out because the overall quality, even the very likelihood of successful completion, depends on it.

Let us consider an example. On the building site observed, the exterior walls of the garages were made up of precast components (columns, and prefabricated panels to be embedded in the columns). One of the teachers of construction present on the building site (we will call him G) explains:

> Our precast panels are going in here (between two posts already in place). You need to know that there's 3 millimetres tolerance each side for their thickness [...] When it's concrete, any lack of care means you have to demolish something.

It is inconceivable to break one of these panels:

> You can't cut a piece that's been worked on.

What about extending it?

> Then you have a joint that's ugly and visible. And the architect may say: I won't accept that.

The practical problems of accurate setting out are unavoidable given the practical problems of manufacturing preformed components which will fit together. In the building industry, official standards define 'tolerances' within which individual components must be manufactured, in order to ensure that they can be assembled successfully with other components[9].

3. THE TRANSFER OF MEASUREMENTS AS A BASIC PRACTICE FOR SETTING OUT

In what follows, we shall concentrate on the practice of transferring measurements during the setting out procedures, from the time the measurements are given to their first being marked out.

3.1. Dimensions in structural drawings

A building project[10] is defined by architects' plans where lengths, angles, differences in height or width, alignment or evenness are represented by dimensions and bench marks. They are themselves explained with the help of a standardised system of drawings. It is customary in the profession, despite the prevalence of SI units (which specify measurement in metres and millimetres), to express measurements in centimetres if they are less than a metre, and in metres to two decimal places if they are above one metre. The tolerances, which are assumed when doing this, are not made explicit: the value of a tolerance is given on the plan only 'if it is necessary to check the dimension or the position, the orientation or the form'.

3.1.1. Determining length by partial and cumulated dimensions

Figure 1 presents an example of an architect's drawing of some building foundations.

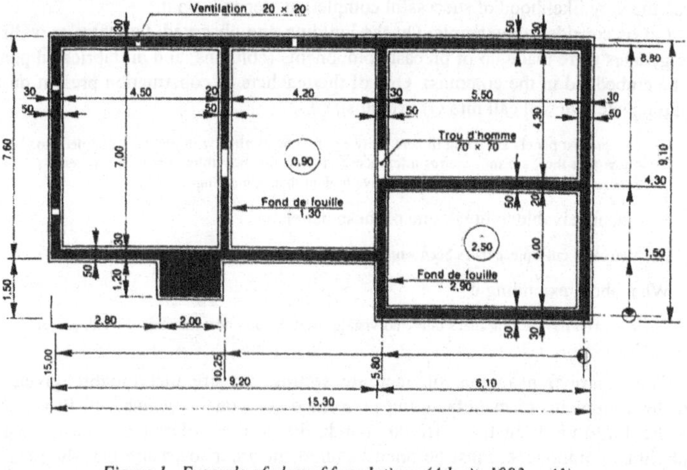

Figure 1. Example of plan of foundations (Adrait, 1993, p.41)

Standard practices dictate where the measurements are placed: for example, measurements inside the drawing describe interior features of the building.

Measurements are written above or next to dimensional lines which are parallel to the building feature; line ends are marked precisely by means of special symbols. Thus, the dimensions relevant to setting out are to be found outside the drawing, laid out along broken lines. The intermediate points of these lines of dimensions are shown by small perpendicular lines - *attachment lines* - referring to lines on the drawing. The latter show features of parts of the construction: centrelines or lines of the walls, for example. There are two types of dimensions, called *partial dimensions* and *cumulated dimensions*. They can be distinguished from one another by the way they are laid out: centred above the dimensional lines for partial dimensions, perpendicular and near to the attachment lines for cumulated dimensions. The three lines at the bottom of figure 1 show this distinction clearly.

In the scale drawing,

-the cumulated dimensions correspond to distances from the same point of origin to successive co-linear points;

-the partial dimensions correspond to line segments - i.e. distances between successive co-linear points.

3.1.2. Dimensions form the intermediate stage between measuring and locating the point at which work is to be carried out.

A simple interpretation of the symbols and marks on the dimensional lines allows us to read them as lines with graduations. One can then use an elementary mathematical model Mo which, from the point of view of the mathematics taught in general secondary school education, relates to the pinpointing[11] of locations on a line with graduations (Bessot & Eberhard, 1983). After choosing some point as the origin, the cumulated dimensions are considered as abscissa and one can pass from one type of dimension to another: from cumulated dimensions to partial dimensions by subtraction, and from partial dimensions to cumulated dimensions by addition. The Mo model establishes a relationship between the two types of dimensioning. But it explains neither the preference for writing cumulated dimensions on some plans nor the importance given to their use on the building site. We explore the preference for cumulated dimensions, and their importance in practical affairs, in the next section.

3.2. Cumulated dimensions, a technique for transferring dimensions from plans to ground

3.2.1. The problem

The basic task consists of transferring several dimensions from the plan, which specifies horizontal dimensions, to a pattern of markers on the ground, in the correct spatial relationships, and using an appropriate scale. The transfer results in pegs being located appropriately on the building plot. The conduct of the building work depends upon these pegs: the problem is to guarantee the quality of the transfer.

3.2.2. The question of mathematical content of dimension transfer techniques
Two techniques for carrying out the task can be distinguished: TCD (Transfer in Cumulated Dimensions) or TPD (Transfer in Partial Dimensions) according to whether the transfer is done in cumulated dimensions or partial dimensions. To return to our initial question: given that the TCD technique can be seen as a technique using the mathematical model Mo, we can ask about the role played by formal mathematical knowledge in the appreciation of qualities which make it preferable to TPD in the eyes of practitioners.

> One of the functions of mathematics is to allow the outcome of an action to be anticipated. The word anticipation encompasses a dual process: prediction and the guarantee of the prediction. (Margolinas 1992, p. 122)

It follows that a number of questions relevant to the practical implementation of the two processes need to be addressed: 'What checks on the accuracy of transfers do TCD and TPD contain? What are the expected margins of error associated with TCD and TPD? What expectations as to the quality of transfers does the choice of TCD include? And finally: on what are these expectations based?

3.2.3. A priori analysis of the two transfer techniques
We have based this on a division of the different components of the techniques into actions and device following the approach introduced by Chevallard (1993):

> the device includes all the elements (tools, materials and processes) which allow a practice to be carried out.

Let us leave for a while the question of the way that various elements come into play in the creation of an alignment; we shall assume the existence of the reference mark M for the alignment, marked by an O on the architect's plan.

-The actions required for transferring horizontal dimensions by cumulated dimensions (TCD).
On the plan, the cumulated dimensions relative to O of each of the relevant points are determined. For this:

if the cumulated dimensions are both related to O and 'attached' to these points, the dimensions are read directly;

otherwise: if the relevant points are all co-linear, on the same line of partial dimensions (and the length of each line segment is known), a number of operations are carried out on the partial dimensions. If the relevant points are not co-linear, the exercise is performed as if everything had to be brought back to the same line[12]. Then the necessary operations may be carried out using different partials or cumulated dimensions (Mo).

In the two latter cases, a list of measurements must be calculated, then noted down on a sketch.

On site, the zero of the measuring tape is placed, vertically from the reference mark M; then the tape is stretched in the direction of the alignment. The pegs are then driven into the ground in succession as the dimensions are read off.

Coordination of the workmen's actions must ensure 'good practice' when using the tape (placing at zero, making sure it is horizontal and pulling it taut), and then that the line is exactly vertical (ensured by competent use of a plumb line).

-Next the spaces between the pegs are checked: the tape is once again pulled taut and the pegs are checked, from the furthest to the nearest.

-Objects forming the device which are the same for TCD and TPD.

These are:

a measuring instrument which is a flexible or metal tape (between 20 and 50 metres long).

an alignment which has been marked out.

Let us consider this second object. First of all, what is it? A material object, with the string pulled taut? Certainly. However, the string on its own has no significance; rather, it is its relationship to other objects, and its location within a conceptual framework which gives the string its significance. Other objects must play a part: the vertical plane which contains it, defined by the direction of the plumb line, and finally the intersection of this plane with the ground, an 'ideal' curve which will be expressed by the marks of the workman or by a line drawn with Cordex[13].

Geometry provides a conceptual structure:

> Determining an alignment. - From the point of view of planimetry, a point on the building site can of course be considered as determined if its projection, that is the vertical line which passes through this point, is known.[_] In the same way the direction of a straight line is of interest in planimetry only by virtue of the vertical plane or alignment which contains it. The alignment is determined by the two *ends* of the straight line. (Hadamard, 1947, p. 287)

Here, Euclidian geometry forms the basis of understanding which allows alignment to be described and the practice of using the string and the plumb line to be justified. It fulfils a theoretical function. On the other hand, the requirement for the 'two ends of the straight line' is *practical* knowledge[14]. Indeed, it is this which empowers the workman when he chooses to attach the string to reliable points of the alignment (which has not yet been marked out) on two setting out pegs. It guides action when checking that the levels of the nails on which the string is to be tied are equal on both setting out pegs: for example, in making a straight horizontal line above the ground. It also ensures that a reference mark M is established on the vertical line of the crossing of two pieces of string stretched between the setting out pegs.

Finally, the transfer of dimensions onto an alignment is described in terms of Euclidian geometry as marking out the distance requirements between straight vertical lines (or parallel planes) in the space designated for building.

-When can TCD and TPD techniques be used?

TCD and TPD can be used to transfer dimensions to any rectilinear line or vertical direction of a flat surface. They play a part in the reciprocal problems of *reading the dimensions* of existing physical components. They play a part also in the space of the building site (50 metre tape) as well as in more restricted spaces.

But the length of the measuring tape limits TCD, which is only applied to lengths which are less than that of the tape, whereas TPD is applied to any length by the reiteration and juxtaposition of transfers.

-Comparative costs of TCD and TPD

Let us suppose that n aligned points are to be marked. TCD requires a single verification of the correct use of the tape. For TPD, which requires n transfers to be juxtaposed, this checking process has to be repeated n times. On the site, TCD is more economical in terms of operations to be carried out than TPD: it is quicker and more reliable, providing fewer opportunities for error when taking measurements with the tape. On the other hand, TCD requires calculations and a preparatory sketch.

-Ways of checking measurements

Let us consider a simple situation:

P indicates a point (on an alignment). The dimensioning provides the partial dimension x relative to the couple of points (P,R).

P is marked, either directly with a single transfer of the length x (TCD), or by n transfers juxtaposed in succession (TPD).

Can *a priori judgements* be made of the confidence intervals for x of the distances of the vertical straight lines passing through R and P, when real measurements are made using either TCD or TPD? In other words, can we estimate the measurement errors inherent in each method?

Guided by the model of the probability theory of errors[15], let us imagine the different causes of error in a transfer as being mutually independent and producing *elementary errors*. Let us also consider errors which are independent of the length of the transfer[16].

Let us suppose that one of them is *systematic*[17].

The properties of the Mo model allow a representation and a calculation to be made. They give a criterion for quantitative comparison:

If E is the error when making a transfer (TCD), the maximum error with TPD is nE.

Let us suppose that there is no systematic partial error (because it is 'corrected', for example). Two models are able to account for *accidental* errors:

M1 model of uncertainties

Working on the hypothesis that the error does not go beyond a given limit E when any transfer is made (uncertainty about a measurement), we have:

$$| eTCD | < E \text{ and } | eTPD | < n E \text{ (uncertainty about a sum)}$$

M2 model of probabilities where the error is seen as a random variable of which the standard deviation is a characteristic. If σ indicates the standard deviation of measurements for any one transfer,

$$\sigma RCC = \sigma \text{ and } \sigma RCP = \sqrt{n}\,\sigma \text{ (rule of 'connected' errors)}$$

In this way, the algebraic theory of uncertainties or the probability theory of errors can work together as a basis for theoretical understanding of good working practices.

For the theoretician: Choosing TCD as opposed to TPD allows errors to be minimised.

For the practitioner:

> It is a cumulated dimensioning which allows the walls and columns of a building to be set out with a minimum of error. (Adrait 1993, p.41)

How is this technique of cumulated dimensions taught and used? An incident from the building site will provide a preliminary answer.

4. AN INCIDENT AT THE CONSTRUCTION SCHOOL

It is the first day at the construction school for a team of pupils who have to set out the prefabricated columns of an exterior wall. The setting out is done in casings. These casings (almost 40cm deep) were created when the concrete for the slabs was poured. They are supposed to bind the columns to the slabs: the columns will be embedded in them and then sealed.

Tutor G explains what is involved:

> They have to place the two columns so that when the crane arrives with the panel (of the wall), they can put it in the right place within a minute (between the columns).

Here, the constraints on the dimensions are closely linked to the constraints of minimising the cost of carrying out the work, as well as to requirements related to the organisation of the work (the crane belongs to the neighbouring building site) and to safety (because of the inherent danger of the manoeuvres).

After checking the position and level of the bottom of the casings (which determines the position of the columns in terms of height), the task being observed consists of marking the position of the base of the columns at the bottom of the casings (horizontal drawing). This is done by marking the location of position templates.

The pupils have a plan with them and they stand around discussing dimensions.

> Pupil 1 (who has the official position of foreman for the day) - We'll have to allow 20 [cm] right there. (They are discussing a partial dimension of the plan for the base of the columns).

The construction teacher responsible for both the students' education and the execution of the work (we will call him R) intervenes forcefully to reprimand the team. He tells them to hurry up.

Questioned[18] by Annie Bessot (A), R and G explain:

> G: They felt the need to whereas they could have worked perfectly well on the existing plan and the cumulated dimensions.

Where there is a question of error

R: It's all cumulated dimension, that. It's obvious that if you transfer the partials, you're going to build up errors all along the way.

And an illustration of the disadvantage of using partial dimensions in the case of a systematic error

G: They have to set out 2 m 50, 2 m 50, 2 m 50, so [this example relates to the setting out done by another team at the same time].

R: To avoid that... a pencil mark, you know, either you make it on the centreline or you make it outside. If you have to repeat it five times, you get to the end with a good centimetre to spare; that's the reason they have to start off in cumulated dimensions.

The teacher describes the actions he was expecting the pupils to carry out: the delay reveals that they have failed to prepare worthwhile calculations from the plan.

R: There you are, they're starting off from an Origin, they're starting off over there, an alignment, and they cumulate everything from that [...] in actual fact you only measure once with the decametre and then you draw. It avoids accumulating errors. So, you need to make a small cumulated plan [...] you can see everything straight away and you can work much more quickly.

On the conditions for using cumulated dimensions

G: By comparison with what we've done in the workshop at school, there are all the real constraints of the building site, so there's an enormous adaptation to be made [...]. At school, they do exercises with '2 metres by 2'. Whereas here, it's 14 metres by 14. So the drawing procedures can't be the same, although the cumulated dimensions, when you're working on 1 metre 50... it's convenient, all the same. (This is correct. The principle of cumulated dimensions is as advantageous for small jobs which demand a high degree of precision as it is for larger ones!)

A: Yes. You can see the advantage of it.

G: It's by far the most economical way of doing it. But you have to tell them, otherwise it would never come naturally.

Actions expected by the teacher in response to a decision required by circumstances which emerge on the building site

Throughout this time, the team is still delaying. One of the pupils finally decides:

Pupil 2: I've got it - we have to do it as a cumulated dimension, and that's it!

A: He seems to know what he wants.

G: He has such control over what he does that all the others follow him [...]. He's the real boss [of the team], the other one hasn't taken any initiative.

For pupil 2, 'we have to do it as a cumulated dimension' is not a response to the need to satisfy the educational aims of the teachers ('What do they expect us to do?'): it is a response to the problem faced by his team. Pupil 2 spotted an opportunity to use cumulated dimensions and for G this recognition appears to constitute a 'professional' course of action.

R's reprimand is occasioned by the delay in carrying out the task. For him, in his position as head of site, the implementation of TCD seems a natural response to his order to set to work: the constraints of measurement tolerance and the need for speed make it necessary. The pupils should have prepared in advance a cumulated plan: they did not do this. The lack of anticipation, their doubts and their delay illustrate the difficulty they may have in recognising an *opportunity to use* TCD. We have here an episode to relate to 'non-routine episodes which [...] involve breakdowns', observed and studied by Noss, Hoyles & Pozzi to make visible mathematics (see article in section 1 of this book).

5. MATHEMATICS AND KNOWLEDGE RELATING TO CUMULATED DIMENSIONS AT SCHOOL

'Cumulated dimensions' are taught for the BT by teachers of 'construction'. They are encountered during the first practical exercises linked to surveying lessons. These classes are structured around tasks of taking measurements of lengths and angles. They aim to inculcate the techniques for using instruments: the pupils learn first and foremost not to make errors[19] and to use instruments with care. A typical exercise from the first chapter of the reference textbook used by teachers consists of sketching, on a sheet of drawing paper, the 'right-angled coordinates' of details of a 'fairly flat' site. A basic procedure taught is to copy out the 'abscissas' from the points of a 'guide line' with TCD and then to transfer them on to the sheet by 'picking them out in cumulated dimensions'[20] along a straight line. The justifications given, 'easy to read and transfer', 'can be transferred to a drawing with speed and precision' are primarily concerned with simplicity of implementation; precision is not justified. TCD is not mentioned again during the rest of the course, but, according to the teachers, it is used several times, in practical exercises or in the workshop.

In BT classes, the actions involved in TCD are an important concern[21] *in teaching*. The teachers justify their advantages in terms of economy and reliability, within the logic of objectives linked to the use of instruments. The *requirements for practical mathematics* in calculating dimensions are officially met[22] the teaching of the school.

Of course the question of *recognising opportunities to use TCD* belongs to the field of didactic perception by teachers. G sets it out clearly. But, from the learners' viewpoint, might the required learning be almost invisible? That may be related to the didactic contract: the use of TCD is present in exercises where it is all but suggested by the information given, either by 'real' plans or by the mode of operating. Indeed, in setting out tasks which are governed by a concern for precision, TCD is implicitly prescribed by the cumulated dimensions (and symbols for their origins) on the plans:

> Theoretically, the [cumulated] dimensions for setting out must figure on all the plans. In practice, they are hardly used except for plans of foundations. (Adrait, 1993, p.41).

In this respect, the situation observed was a kind of breaking of the didactic contract with the team of pupils: first day on the building site, partial dimensions on the initial plans, teachers who encourage problem-solving by leaving students to sort themselves out, in this case simply take over important decisions:

> Pupil : We're trying to sort things out. When we have a problem. We're here to learn. But how does the teacher usually teach on the site?
>
> G: When you have responsibility for 15 pupils with as many technical problems, there are times when you don't have a choice: you have to show them, do it, and explain afterwards.

How might cumulated dimensions be explained? In BT classes, the notion of error restricts the reasons for choosing TCD to the simple imperative of maximising efficiency of operating. Indeed, the teachers of 'construction' do not have to take responsibility for theoretical knowledge of errors. Teaching about errors is to be found in the 'Building' section of BTS[23] in the surveying coursebook. It is presented as a 'structured whole' around a typology of errors of measurement. Notions of precision and accuracy are defined and studied, together with the concepts of systematic and accidental error. A (probability) theory of accidental error is presented: the mathematics surrounding the notion of error are developed with reference to a whole corpus of measurement situations known to the pupils, and for which they establish a model (of the M2 type).

On the other hand, the objectives of the BT lead to errors being regarded as failings. Systematic errors giving rise to gaps *not accepted by the teacher* should be searched out and corrected - which is the reason they were put forward in R's justification of the choice of TCD (as against TPD). The property of being able to minimise systematic error is shown by an example and a shift in meaning: 'errors build up' refers to the reiterated addition established by the elementary Mo model.

For R, minimisation of the systematic error based on school mathematics is therefore self-evident. One can make the assumption that it is the same for minimising the 'uncertainties'. But R that cannot formulate this property because it would require the notion of *absolute error*, which is absent at this level.

6. CONCLUSION

In fact, the general teaching of mathematics today (whether in secondary school or high school) does not include any reference to the notion of error nor, a fortiori, to questions of errors linked to measurement. The notions of uncertainty, of absolute and relative errors, which were still present in high school education in the 1960s, have disappeared from the curriculum. The elements which allow the notion of absolute error to exist are present, but they are set in the context of numeric approximation and have no formal institutional link with the problem of error measurement. The algebraic knowledge necessary to justify the importance of cumulated dimensions by the model of uncertainties is taught in high schools. This knowledge could then take the role of basic knowledge. But this role is very unlikely in technological education (in courses of building construction) at BT level, given that the mathematicians have not taken on responsibility to teach it and that moreover a simple practical justification exists.

This absence of the notion of error in the curriculum may be set alongside a general trend for avoiding problems of actual measurement in school education in France (Brousseau, 1992). This phenomenon goes against the condition proposed by Gillespie in section 1 of this book, for the successful integration of mathematics in professional teaching.

Thus, the teaching of cumulated dimensions in BT courses reveals a *gap* as to the status of mathematical knowledge which is 'useful' for vocational training. In this establishment, for the practices proposed, working knowledge of algebra is elementary knowledge, taught beforehand. 'Intermediate' knowledge of algebra, which is taught at the beginning of high school, could function as basic knowledge, (before a perspective of probability is introduced[24]). Due to there being no questions about errors of measurement in mathematics, these are 'outside the boundaries', even though they could be taught at BT level, as fruitful 'points of articulation between school maths and outside ('target') activities'. (See the article by Evans in section 1 of this book).

These days, teaching mathematics in vocational training establishments is not a trying experience. Teachers are questioning the relevance and role of mathematical knowledge, both for the present and the future. The failures to connect mathematical theory to professional practice revealed in the short study we have presented perhaps goes some way towards explaining this feeling of unease.

7. REFERENCES

Adrait, R., and Sommier, D. (1979). *Le Guide du Constructeur en Bâtiment.* Paris: Hachette.

Artaud, M. (1993). *La Mathématisation en Économie Comme Problème Didactique, Une Étude Exploratoire.* University of Aix-Marseilles II.

Bessot, A., and Eberhard, M. (1983). Une Approche Didactique des Problèmes de la Mesure. *Recherches en Didactique des Mathématiques, 4*(3), 293-324.

Bessot, A., and Eberhard, M. (1995). Le Problème de la Pertinence des Savoirs Mathématiques Pour la Formation aux Métiers du Bâtiment. In G Arsac, J. Grea, D. Grenier and A. Tiberghien, (Eds.), *Différents Types de Savoir et Leur Articulation.* Grenoble: La Pensée Sauvage.

Borel, E., Deltheil, R., and Huron, R. (1962). *Probabilités, Erreurs.* Paris: Librairie Armand Colin.

Brousseau, G. (1980). Les Échecs Électifs dans l'Enseignement des Mathématiques à l'École Élémentaire. *Revue de Laryngologie Otologie, Rhinologie, 101*(3-4), 107-131.

Brousseau, G., and Brousseau, N. (1992). Le Poids d'un Récipient. Étude des Problèmes de Mesurage en CM. *Grand N, 50* (Published by IREM of Grenoble), 65-87.

Chevallard, Y. (1991). *La Transposition Didactique - du Savoir Savant au Savoir Enseignè, (2nd edition),* Grenoble: La Pensée Sauvage.

Chevallard, Y. (1991). *Dimension Instrumentale, Dimension Sémiotique de l'Activité Mathématique.* Paper presented at the Seminaire de Didactique des Mathematiques et de l'Informatique 1991-1992, LSD2-IMAG Laboratory, Grenoble.

Hadamard, J. (1991). *Leçons de Géométrie. II Géométrie dans l'Espace.* Paris: Jacques Gabay.

Margolinas, C. (1992). Eléments pour l'Analyse du Rôle du Maître. Les Phases de Conclusion. *Recherches en Didactique des Mathématiques, 12*(1), 113-158.

Mercier, A. (1995a) La Biographie Didactique d'un Élève et les Contraintes Temporelles de l'Enseignement. *Recherches en Didactique des Mathématiques, 15*(1), 97-142.

Mercier, A. (1995b) Le Traitement Public d'Elements Prives du Rapport des Élèves aux Objets de Savoir Mathématiques. In G. Arsac, J. Grea, D. Grenier and A. Tiberghien, (Eds.), *Différents Types de Savoir et Leur Articulation*. Grenoble: La Pensée Sauvage.

Nathan. (1994). *Précis de Chantier, Matériel et Matériaux, Mise en Oeuvre, Normalisation*. Nathan.

Olivier, E. (1976). *Implantations, Tracés, Nivellement, Relevés : Travaux Pratiques*. Paris: Entreprise Moderne d'Edition.

Renaud, H., and Leterte, F. (1978). *Ouvrages en Béton Armé*. Paris: Foucher.

Séré, M. G., Journeaux, R., and Larcher, C. (1993). Learning the Statistical Analysis of Measurement Errors. *International Journal of Science Education, 15*(4), 427-438.

8. NOTES

[1] this article is based on research carried out by the following team: Bessot, Déprez, Eberhard.

[2] Lycée Technique et Professionnel Roger Deschaux, Sassenage, in the suburbs of Grenoble, France.

[3] Two years of study in a technical high school (lycée technique, LT) are required for the Brevet de Technicien (BT), which is a final examination of the same level as the baccalauréat leading to middle management jobs in building and public works.

[4] Theoretical tool introduced by Artaud (1993) to analyse the function of mathematical knowledge in economics. See also Mercier elsewhere in this book.

[5] The didactic contract is defined by Brousseau (1980) as being 'the set of (specific) actions [relating to the subject-matter taught] which are expected by the pupil of the teacher, and the set of actions expected by the teacher of the pupil'.

[6] See the article by Bessot, in section 3 of this book, for a general study of the different practicalities of reading plans.

[7] This reference work is intended for practitioners on the construction site, future technicians in the profession, and teachers.

[8] The adjective 'exact' may be explained by the author's choice to refer systematically to a geometrical model. In Nathan (1994), a reference work for defining tolerances, one reads: 'Because of the imprecision of means for carrying work out, no job or part of a job can be executed to exact dimensions'.

[9] Tolerance first came into being because of the problem of adjusting components in mechanics. In building, the present-day importance of prefabrication has led to its function being modified: the rules of the art of tolerance are tending to be incorporated in the lists of official standards.

[10] This description is based on Adrait (1993), a work intended as a reference for a very wide reading public (high school pupils, students in higher technical education, professionals). Because of this, it represents the technical and statutory norms of the profession

[11] This pinpointing of locations is confined to the handling of positive numbers if one of the furthest marks is chosen as starting point, as is generally the case on plans (horizontal dimensions). Negative numbers may play a part with the dimensioning of levels by reference to an origin set at ground level, when using cross sections.

[12] Which will possibly be checked by the notion of orthographic projection. See the article by Bessot, section 3 of this book.

[13] Cordex (trade name) is a length of string coated with coloured powder. When taut, and pulled from a centre point then released, the coloured powder is deposited in a straight line.

[14] This dual function for geometrical knowledge (taught before this level) is to be found in the 'construction' teaching for BT

[15] One might refer, for example, to the treatise of the mathematician Borel (1962).

[16] The use of the steel tape can lead to corrections of reading according to length. In the absence of these, the hypothesis might be referred to relative errors.

[17] It is reproduced 'identically' (same degree of severity, same manner of making the error) if the transfer is begun again by the same method.

[18] The observational method is described in the article by Bessot in section 4 of this book.

[19] The zero position, with 0 which are not placed identically, is the cause of recurrent errors.

[20] The use of a standardised scale (1/50 or 1/100) allows people to rely on units whilst possibly only multiplying by 2 (10 m make 20 cm etc.).

[21] After the event, when filling in a questionnaire, R remembered a problem which had arisen in the construction school: 'for the alignment of a wall separating the garages, a discrepancy was noticed: some cumulated dimensions had been wrongly transferred'.

[22] When 8 out of 29 technology teachers from the Deschaux high school, in reply to the question: 'what are the main mathematical concepts your pupils should know?' state 'the four basic operations', it begs the question of whether these are learnt.

[23] Brevet de Technicien Supérieur, a post-baccalauréat qualification which takes over two years to prepare.

[24] See Séré (1993) for the case of physics.

[illegible] The use of a [illegible] scale [illegible] allows people to rely on [illegible] whilst possibly only multiplying by [illegible]

[illegible] After the event, when filling in a questionnaire, K remembered a problem which had arisen in the [illegible] the measurement of a wall separating the garage [illegible] the completed operations had been wrongly transferred.

[illegible] When [illegible] from the [illegible] in reply to the question: "What [illegible] mathematical [illegible] your pupils should know", states the [illegible], it is a question of what they need to know.

[illegible] Except in French-speaking countries, a [illegible] qualification which takes over two years to prepare.

[illegible] See [illegible] for the case of [illegible]

CHAPTER 4

THE INTEGRATION OF MATHEMATICS INTO VOCATIONAL COURSES

SOME ISSUES AND CONCERNS

JOHN GILLESPIE

Abstract and introduction. Some college-based vocational courses for post-16 students in England require them to demonstrate proficiency in the application of mathematical techniques in addition to meeting their main course requirements. A view is held that this is most meaningful when it can be integrated into the vocational work. But this is not easy to carry out, particularly in courses where mathematics does not play a major role. This chapter discusses some of the issues and suggests courses of action that could enable integration to be more effective. Examples are taken from classroom observations and interviews with students.

1. MATHEMATICS WITHIN VOCATIONAL EDUCATION IN ENGLAND AND WALES

Currently in England, each year at least 20% of school students reaching the minimum school leaving age of 16 move on to vocationally-oriented education courses leading to General National Vocational Education (GNVQ) awards. For the most part, these courses are college based.

The GNVQ courses lead to awards at three levels:

-Foundation or level 1 - normally based on one year courses, suitable for students typically with few General Certificate of Secondary Education (age 16 school leaving qualifications) at low levels

-Intermediate or level 2 - normally based on one year courses, suitable for students with some GCSEs

-Advanced or level 3 - normally based on two year courses, suitable for students with a range of GCSEs including some at higher grades.

GNVQ courses in Business, Health and Social Care, Art and Design, and Leisure and Tourism have proved popular at all levels.

Other vocational training is provided in England and Wales for post 16 young people who move directly into paid employment (through National Trainee and Modern Apprentice schemes). These are outside the scope of this chapter.

Bessot & Ridgway (eds.), Education for Mathematics in the Workplace, 53—64.

2. KEY SKILLS – INCLUDING APPLICATION OF NUMBER

A feature of GNVQ awards is that, as well as showing competence in the vocational areas, students are required to show competence in three 'Key Skills' (that is, 'generic' or 'core' skills) of Application of Number, Communication and Use of Information Technology. Application of Number - the application and use of mathematics in the context of other work - has developed from and includes earlier understandings of 'numeracy', such as

> an ability to make use of mathematical skills [...] to cope with the practical demands of (the individual's) everyday life [...] and to have some appreciation and understanding of information presented in mathematical terms [...] for instance in graphs, charts, percentages (Cockcroft 1982, para. 39)

A major objective is that students should develop and demonstrate these key skills in the context of their vocational work - that is, integrated into the vocational programmes and assessment. This objective is an outcome of a widely-held dissatisfaction with both the levels of competence in techniques in the three key skill areas among those seeking to enter employment, and the perception that even individuals who possess some technical expertise in these areas (for instance, as evidenced by their GCSE qualifications) are not necessarily able to apply these skills effectively in context (Cockcroft 1982).

For Application of Number, in practice this objective has proved difficult to meet in vocational areas where mathematics does not have an obvious major role.

In this chapter I want to look at some practical issues affecting

-the more effective development of Application of Number skills and

-their integration into some GNVQ Intermediate level programmes

I will illustrate the issues with examples from the project 'Mathematical Skills in GNVQs' (Gillespie 1999) which investigated these issues in the contexts of two Intermediate level GNVQ courses - in Health and Social Care and in Art and Design. (These courses could constitute 'practices' as proposed by Evans, q.v.). The examples I believe illustrate a pragmatic approach on the part of the students, where 'occupational or professional concerns take precedence over those that are mathematical' (Noss, Hoyles and Pozzi q.v.). The Health and Social Care examples illustrate the development of reflection on and questioning of previously accepted views, whereas the Art and Design examples are characterised by I believe a problem solving agenda. But it would have been possible to choose other examples from each area to show both these agendas (for instance, surveys on people's perceptions of pleasing proportions and the Golden Ratio in another Art and Design programme).

Many of the students observed during the project came to the courses with only modest success at school, and with below average GCSE grades. For them, the vocational courses opened doors of success and interest which, from interview feedback, were the exception during their school days.

3. WAYS OF PROVIDING SUPPORT FOR APPLICATION OF NUMBER

A majority of vocational tutors interviewed during the project saw the requirement to incorporate 'Application of Number Skills' as an imposition, and one which they did not feel able to satisfy. Often they turned to specialist tutors with a maths background to meet the need.

One response was to provide 'stand-alone' maths techniques courses often outside specific vocational contexts. But many students saw such courses as not relevant to their main courses, and there was little evidence of their being able to transfer skills from them to their vocational work.

What appeared more successful was where small teams of staff were able to work together, with a maths specialist providing basic mathematics support teaching while also being aware of the content of vocational courses and keeping in touch with vocational teaching colleagues. But the logistics of such co-operative work could be difficult. In addition there were often significant differences in perspectives of vocational and maths specialists which needed to be worked through. Colleagues from each 'culture' had to have time to work together, and with the same students, on common work. When this took place, it appeared that shared understandings grew, with preconceptions changing to appreciation of others' viewpoints.

4. VIEWS FROM TWO CULTURES

These perspectives emerged from observation and informal discussions with both mathematics and vocational staff during the project.

Within the culture of the mathematics classroom, mathematical thinking is 'comfortable' and the norm; it is the focus for work; in such an environment it has been customary to 'see the maths' in situations. The application of mathematics has typically involved seeking out situations in which specific pieces of maths can be demonstrated and developed. The point of the exercise is in the exploration and solution of a maths-centred problem, or the development of mathematical thinking and techniques. Problems are contexts for doing maths.

In contrast, within some vocational cultures, both staff and students can see mathematics as peripheral to their work, admit to feeling threatened by and hostile to mathematical thinking and have a limited view of opportunities it can offer. These feelings seem often based on the individual's own personal negative and narrow experiences of maths from their own school days. The offer of support from a maths specialist who is not aware of these feelings can simply serve to reactivate them and inhibit co-operation. Further problems arise if the maths specialist is not willing to take time to understand the aims and perspectives of the vocational course.

These two cultures I believe relate directly to the two approaches to knowledge outlined by Eberhard (q.v.). In the examples below the students respond to the requirements of the task in hand or their new insights, rather than to what they believe 'the teacher expects them to do ('What do they expect us to do?' – Eberhard referring to an incident in the construction school).

5. ADDING VALUE TO VOCATIONAL WORK BY INTEGRATING MATHEMATICS INTO IT

From the vocational viewpoint, application of mathematics only has a purpose if it 'adds value' to the course - where there is a point in doing it. For instance, there is no point in carrying out a statistical survey for its own sake - but if survey is a means to a vocational end, say as part of a health and social care study, then the survey has a purpose. Once the purpose of the survey is clear, maths-based skills can add insight and doing the maths changes from being a chore to opening up opportunities. The discussion sheet 'We can carry out a survey' (fig. 1), devised as a result of experiences within the project, has prompted some of these issues and opportunities.

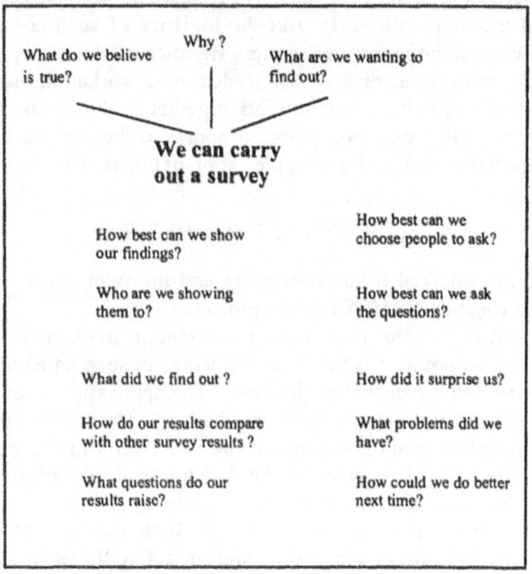

Figure 1. A survey should be a means to an end, rather than an end in itself

6. TAKING CHARGE OF LEARNING

Integration means that mathematics is undertaken to improve the quality of work of vocational work, not the other way around. When some students felt that maths was aiding a vocational end, and that they were in some sense in charge of their learning, using the maths specialist as a resource, they were prepared to undertake work that would have seemed artificial at school.

Student A It was better than school. It was only doing what you needed to know more about than having to go through everything.

Student B Yes it's not like school where you have to do it in an hour and that was it...

Student C Doing Application of Number in Health and Social Care, you don't treat it as maths.

Interviewer *How does that make it different then?*

Student C It makes it more interesting - it's fine if it is part of our work

A characteristic of integration is that the individual is aware of selecting and using mathematics techniques as tools to help meet the need of the (vocational) task in hand, prompting the self-diagnosing of some weakness (e.g. 'I need to draw a pie chart, so I need to find out how to do it').

7. FORMING HYPOTHESES

An effective use of quantitative data and calculations can be to correct preconceptions and sharpen perceptions. This requires that students commit themselves in advance to what they think will be the outcome of, for example, a survey on a particular topic. Rather than just accepting the findings as so much more data, by contrasting survey results with their expectations, the students find themselves questioning their original beliefs, seeking reasons to explain the differences - hence leading to new understandings (see examples 1 and 2 below).

8. INTEGRATING OR EMBEDDING MATHEMATICAL SKILLS

Mathematical skills can also be used without the user being conscious of their presence, where the applied mathematical skill is 'embedded' in the activity. This lack of consciousness may make it harder for the user to develop or refine the skill or use mathematical reasoning to correct mistakes - in contrast the individual who is aware of integrating a particular skill and may be in a better position to modify its use, or select an alternative better approach.

The distinctions between embedding and integrating generic skills are evident in higher level programmes too. Hodgkinson (1996) gives an authoritative account.

9. PROJECT EXAMPLES

I turn now to some examples from the project, with commentaries, to illustrate what appear to be some effective approaches for improving the integration of Application of Number within GNVQ courses. Despite these approaches, many difficulties remain - the examples show this also.

9.1. Examples from Health and Social Care

9.1.1. Example 1

In this example the vocational requirement was to 'Explore how inter-personal relationships may be affected by discriminatory behaviour'.

As part of their work in Health and Social Care, one group of students took the issue of bullying as a theme for an assignment, under the heading 'It's not fair'. This provided opportunities for them to draw on and apply simple mathematical concepts and ideas to do with devising, collecting and displaying quantitative survey data They carried out a survey as a means to investigate their area of interest, in contrast to carrying out a survey as an end in itself (one thinks of surveys of questionable justification undertaken in maths classes, e.g. of colours or makes of cars). Here, however, the application of mathematical ideas helped to students to gain insights into their main area of interest.

Each student devised an interview schedule, obtained data from twenty people and reported on their findings. Many of the students had experienced being bullied themselves, but saw themselves as exceptions. They came to the assignment with views of what they thought they would find - in other words they had formulated hypotheses to test.

It was the contrast between what they thought they would find and what they actually found from their surveys that brought the activity alive.

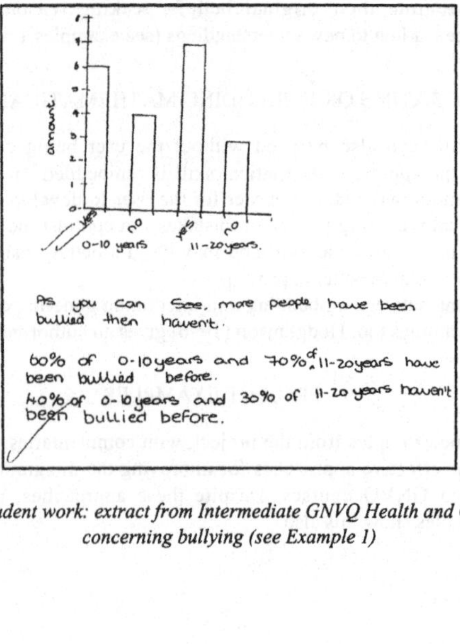

Figure 2. Student work: extract from Intermediate GNVQ Health and Care project concerning bullying (see Example 1)

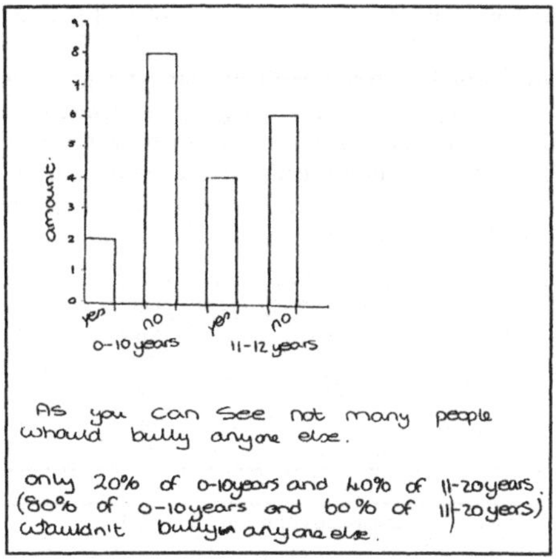

Figure 3. Student work: extract from Intermediate GNVQ Health and Care project concerning bullying (see Example 1)

For instance, in answer to *Were you surprised by how it turned out?*

> student D: Yes, I found that more people nowadays are getting bullied because of things they wear, like not (having) designer labels, or you don't wear designer clothes... When my mum was the same age as us it was because they were a girl and they wanted to do things (not considered appropriate for girls), or they wore their hair in pigtails.
>
> student E: I thought about one in five people would have been bullied, but in fact most people I asked had been bullied (see figs 2 and 3)
>
> student F: *and how about who did the bullying?* It was mostly girls towards the girls - I was quite surprised about that.
>
> (from student work, Gillespie 1999)

The mathematics involved with the surveys was simple, but the effect of the surveys was to give the students involved new insights.

9.1.2. Example 2

In this example the vocational requirement: was to 'Investigate personal health; present advice on health and well being to others'

Body-mass charts and tables, giving information on desirable weight ranges for given heights, are included in many courses. These provide relevant examples for developing graph and table reading skills.

But there are also opportunities here to put forward hypotheses and challenge beliefs about perceptions and realities. For instance, extracts from nationally-collected data - such as that obtained from interviews with 4436 young adults in the spring of 1990 (Rudat 1992), see fig 4 - form starting points, to raise questions and to prompt comparisons with local data collected by students among friends and colleagues.

Here again, by calling on (and probably strengthening) their abilities to understand tables of data including percentages, the students were able to gain insights into their main area of concern – in this case perceptions of body-mass. Again, by integrating their required mathematical work into their main area of interest, the mathematics was seen as an essential tool, rather than as an area of academic curriculum of little relevance.

Body mass index: by age and gender			
	Overweight %	Normal %	Underweight %
All	9	56	35
Women 16 - 19	8	52	40
Men 16 - 19	9	60	31

Perceptions of weight: by gender			
	Overweight %	Normal %	Underweight %
All	29	52	19
Women 16 - 19	40	48	12
Men 16 - 19	19	56	25

Figure 4. Survey data and perceptions of body weight (see Example 2)
Source: Rudat 1992

Comparisons show that 'although women are more likely to be underweight than men, twice as many women think of themselves as overweight than men'. Other equally telling contrasts between reality and perceptions are evident. Such comparisons provide the basis for hypotheses about young people's beliefs about themselves, for instance, which can then be tested in a small way by members of the Health and Social care group through their own surveys.

9.2. Examples from Art and Design

Several examples of mathematical understanding and insights were evident in Art and Design courses - to do with estimation, precision of measurement, scaling and proportion. Some of the students had a narrow, arithmetic oriented view of what constituted mathematics and saw what they were doing as 'not mathematics'. Typically the feedback was visual rather than quantitative (e.g. inappropriate location of vanishing points in perspective, body parts out of proportion). By and large these students did not draw on mathematical ideas or techniques to correct errors - their mathematical thinking was too deeply buried. Others were aware that they were integrating simple mathematical techniques into their work, and could bring them to bear in improving the quality of their work.

9.2.1. Example 3

In one course, students were designing and producing op-art pictures. Some of their designs are shown in figure 5.

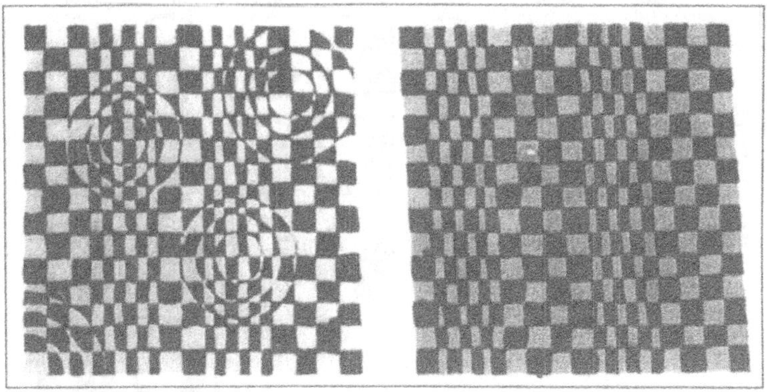

Figure 5. Student work: extract from Intermediate GNVQ Art and Design op-art work (see Example 3)

> Student G It's meant to go along in and then back out....I should have taken more time over spacing squares and sighting them- so actually gradated down in size more. What I have got here I start off with a centimetre and a half squares, they stay the same too late, they don't change down small enough in time...

G is aware of the shortcomings of his design, and though unaware at this stage of any mathematical technique which would help him achieve the effect he wants, he is clear that precision in measurement is critical to achieving a satisfying result.

As with the examples in Health and Social Care, the student is calling on simple mathematical ideas and skills to improve the quality of his work. He sees that precision in measurement brings gains in quality. At the same time ideas of consciously non-linear variation are developing

9.2.2. Example 4

In another student's work, where the requirement of the exercise was to design a letter form based on the square, circle and triangle, the care taken in setting out the designs and the precision of execution contributes directly to the effectiveness of the product (figure 6).

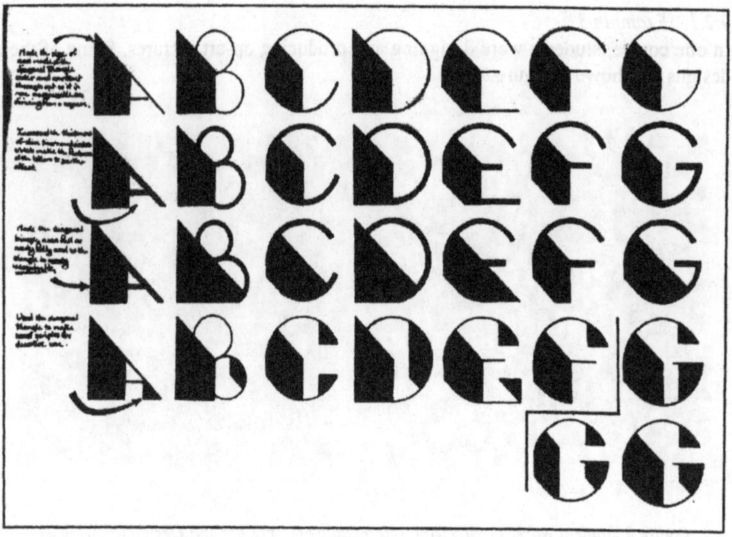

Figure 6. Student work: extract from Intermediate GNVQ Art and Design letter form design work (see Example 4)

9.2.3. Example 5

A further project required the students to draw part of a complex building which was not accessible for measuring. A student selected a particular viewing point and, by holding a pencil or ruler at arm's length and lining it up with part of the building, was able to build up a collection of rough measurements which could then be scaled up or down to serve as aids in producing a final drawing of the building as seen from that viewing point, to any required size. Figure 7 shows one result.

Figure 7. Student work: extract from Intermediate GNVQ Art and Design architectural sketch work (see Example 5).

Here ideas of scale and proportion are central to what the student is attempting. Classroom exercises of scaling geometric shapes up and down come alive as the student really depends on the correctness of his rough measurements to help him produce an effective and rewarding sketch.

As with the Health and Social Care examples, the Art and Design students had access to Application of Number support from a key skills tutor who had an appropriate vocational background. Not only was he able to support students when requested in their vocational art work, but he also provided Art and Design based mini-projects to enable students to develop their Application of number skills directly.

10. CONCLUSION AND SUGGESTIONS

To conclude, the five examples illustrate how 'Application of Number skills (that is application of mathematical ideas and skills), which students are required to demonstrate in order to gain GNVQ qualifications, came come alive when they are integrated into their main areas of interest and concern. Students gain power – in skills – in insights – in reasoning – in quality of work – through selecting and applying simple mathematical ideas. The necessity of mathematics learning becomes a virtue.

Suggestions

Based on these examples and others noted as part of the project, the following suggestions are offered:

-Many students and their vocational tutors are likely to have a negative view of the value and relevance of mathematical skills in their areas. This may be a consequence of their own negative experiences with mathematics.

-Students (naturally) see their vocational work as their central concern. So they see value in the integration of communication, number and other skills with this work, if the skills add value to it. But this is often hard to achieve, in part because of the negative and limited view of mathematics held by some vocational tutors.

-Students are prepared to develop particular number and other maths skills to meet specific vocational needs.

-There are opportunities for raising the quality of insights in vocational work, for example through the use of hypothesis posing or reflecting, which can be enhanced by the use of simple number skills.

-There may be opportunities for raising the quality of students' work and understanding through more creative use of mathematical skills, but this requires a sympathetic understanding of the vocational goals and settings on the part of specialist support staff.

11. REFERENCES

Cockcroft, W. H. (1982). Mathematics Counts: Report of the Committee of Enquiry. London: HMSO.

Gillespie, J. A. (in press). Mathematical Skills and GNVQs.: Project Report to the Nuffield Foundation.

Hodgkinson. (1996). Changing the Higher Education Curriculum - Towards a Systematic Approach to Skills Development. London: The Open University and Department for Education and Employment.

Rudat, K., Ryan, H., and Speed, M. (1992). Today's Young Adults - 16-19 Year-olds Look at Diet, Alcohol, Smoking, Drugs and Sexual Behaviour. London: Health Education Authority.

CHAPTER 5

MATHEMATICAL MEANS AND MODELS FROM VOCATIONAL CONTEXTS

A GERMAN PERSPECTIVE

RUDOLF STRAESSER

Abstract. The paper starts with a description of vocational education in Germany and the teaching/learning of mathematics within this type of training/education. It analyses the relation between a modelling approach and legitimate peripheral participation at work and in a classroom environment, and shows the utmost importance of mathematical (conceptual / algorithmic) and machine-based means when coping with vocational contexts. The paper ends with speculating on the future of teaching and learning mathematics not only in vocational contexts.

1. VOCATIONAL EDUCATION IN GERMANY: THE ORGANISATION

Training and education in technical and vocational education in the Federal Republic of Germany (FRG) can be divided into two main categories: There is part-time vocational training and education for the vast majority of young people in technical and vocational education (about 40% of the population aged 16 to 18), and full-time vocational (classroom-type) education for a minority (about 15%). In addition to vocational education, Germany had a certain renommee for his 'general' education in Gymnasium; which takes around 25% of the age group between 16 and 18 (for details cf. Bundesministerium 1997, p. 26f). In addition to that, a whole variety of institutions (from public to private ones) offer training and education for adults in different types of organisation (from computer based self learning units to year long qualification and training in evening or full-time courses).

The part-time vocational training and education normally consists of classroom-type schooling one or two days per week complemented by three or four days of vocational training in companies (the West German 'dual system' of vocational training). This type of initial vocational training normally lasts three years and offers a vocational certificate as a qualified worker (*Facharbeiter*) to the successful student. In the classroom part of the training, in the *Berufsschule*, vocational arithmetic (*Fachrechnen*) is usually taught for two or three hours per week with the aim of giving a numerical foundation and interpretation of vocational phenomena,

This paper was first published as Straesser, R. (1999). Mathematical Means and Models for Vocational Contexts - A German Perspective. In: Baldwin & Roberts (Eds.): Mathematics - the next millenium. 17th Biennial Conference of AAMT, Adelaide, Australian Association of Mathematics Teachers Inc. (AAMT), 17-29. (ISBN-no.: 1 875900 33 0) underpinning vocational knowledge with an arithmetical analysis ('zahlenmäßige Deutung und Durchdringung von beruflichen Erscheinungen'; see Wolff 1958; Straesser 1982).

Full-time vocational education takes place under a variety of organisational forms with a variety of aims and places in the student's career. There is one major distinction in full-time vocational education, namely the line between prevocational training after general education and 'higher' vocational training. Prevocational training and education lasts for one or two years, normally preparing the student for a whole range of related professions (such as commerce, metalwork, or social vocations) and usually or legally offers the possibility to shorten the time spent in the dual system mentioned above. Higher professional training prepares students to upgrade their professional competencies or to enter an academic career. Mathematics is a subject in all these types of colleges and has the dual goal of repeating the mathematics of the preceding general education and giving the student an introduction to higher mathematics (e.g., calculus) to ease the transition to university education.

2. TEACHING MATHEMATICS IN GERMAN VOCATIONAL EDUCATION

2.1. *Aims of mathematics education and training*

In general as in vocational and technical education, the teaching of mathematics can have a broad variety of aims that are related to the overall aims of education and training (cf. Blum&Straesser 1992):

1. Pragmatic aims: Mathematics teaching is intended to help students to describe and understand relevant aspects of extra-mathematical areas and situations and to cope better with them. Such situations may originate from:

1a. present or future fields of study or profession, or from other vocational subjects at school, or

1b. daily life and the world around us, or from other general subjects at school.

2. *Formative aims:* By being concerned with mathematics, students should:

2a. acquire general competencies; for instance, the ability to argue, to solve problems, or to translate between the real world and mathematics, as well as attitudes such as being comfortable when facing new problem situations and being willing to make intellectual efforts, and

2b. obtain pleasure and enjoy their activities.

3. Cultural aims: Students should be taught mathematical topics:

3a. as a source of philosophical and epistemological reflection, including individual and human self-reflection,

3b. to generate a balanced, comprehensive picture of mathematics as a science and part of human history and culture, including a critical appreciation of actual uses (and misuses) of mathematics in society, and

3c. to produce knowledge, skills, and abilities related to special mathematical themes, and to create a 'meta-knowledge' of mathematics.

At a first glance, only aim 1a seems to be vocational in a narrow sense - consequently vocational use of mathematics should be the focus of mathematics teaching in vocational education. All the other aims seem to belong to general education. In addition to that, there are also good indications that even specific mathematical competencies can only be helpful and be used in the workplace if 'transfer'-competencies (as identified in aim 2a) are at hand - so the teaching of mathematics should enhance the transfer of mathematical skills, techniques and concepts over different situations.

Information on the use of mathematics in the workplace shows the importance of elementary calculations with numbers and technical magnitudes (including percent and the rule of three or proportion) and the use of tables and diagrams. For technical vocations, basic algebra and geometry are widely used. At present, the growing use of computers and related organisational changes seem to be inducing a growing need for geometry in domains like metalwork, electricity, and architecture (especially in technical drawing or CAD and with CNC-machines). In business and administration some basic ideas from statistics seem to be relevant.

2.2. Mathematics education in part-time vocational colleges

In West German part-time vocational colleges, mathematics teaching aims at helping students describe, understand, and master special vocational situations and subjects. It concentrates on elementary calculations with numbers and technical magnitudes and on the use of tables and diagrams. For technical vocations, basic algebra (such as simple equations, formulas, and functions) is taught. Two - and (especially for metalwork and architecture:) three-dimensional-geometry are normally taught to the students or apprentices in specific technical drawing courses. Business and administration courses generally concentrate on a large variety of calculations of interest (like discount, bank drafts, bills, credit and loans, insurance). Some basic ideas from statistics play a specific role in the insurance business (cf. Bardy et al. 1985, pp. 49-63). The majority of these topics also belong to the junior secondary curriculum and should simply be repeated in vocational colleges. Nevertheless, these topics have to be treated again in vocational education for a number of reasons: apart from the time lag between general and vocational education, general education seems to specifically neglect the use of technical or physical formulas as well as three-dimensional geometry.

In vocational colleges, these topics are normally taught in close relation to vocational and professional knowledge, and nearly no disciplinary mathematical structure is followed. Mathematics is taught and used as a mere tool and by-product of vocational knowledge. With an orientation reduced to Aim 1a, the teaching method normally follows three steps: the topic is introduced (or repeated) with the help of some model examples ('problems'). The solution process is condensed to hints and algorithms (often put into formal instructions like 'keep in mind that ...'). Finally, the algorithm is inculcated by solving a number of schematised problems having the same structure (cf. Bardy et al. 1985, pp. 64-71). Reasons for the algorithms are normally not given. The solution process is reduced to the application of recipes to vocational and professional problems. There is nearly no discussion of strategies to link vocational and professional situations to mathematical concepts. As a consequence, differences in the use of mathematical concepts creep in with no legitimisation (e.g., the widespread use of algebraic formulas in technical vocations as opposed to the relative neglect of this topic in business and social vocations).

Apart from the dominance of Aim 1a in part-time vocational colleges, there are additional reasons for the teaching style described above. Teachers try to adjust their teaching style to the motives and competencies they expect from their students and they think students or apprentices in part-time vocational colleges are not as interested in theory and intellectually oriented as university-bound students of the same age. Standardised final examinations (from extra-college bodies) also play a role in the definition of topics and teaching style. On the other hand, detailed analyses of the final mathematics examinations clearly show that the most important source of errors and mistakes are those coming from difficulties with finding the appropriate mathematical model for the problem to be solved. This is a clear indication of the fundamental role of Aim 2a, especially the ability to translate between the real world and mathematics, thus playing down the role of recipes and algorithms for tomorrow's qualified worker.

In recent years, an additional trend can be seen especially for highly automated vocations (like parts of metalwork, electricity, electronics and business and administration): Separate teaching of mathematics apart from vocational subjects seems to be reduced and is - by means of curricula - officially integrated into the teaching/learning of vocational problems. Consequently, mathematics is drowned in vocational contexts and seems to vanish from the vocational training and education. Ambivalent reactions from teachers and students are heard about this development.

2.3. Mathematics education in full-time vocational colleges

The mathematics teaching in German full-time vocational colleges often imitates the mathematics teaching of comparable general education (especially the Gymnasium). 'General' competencies (especially the development of logical thinking, cf. aim 2a) are at the heart of the curriculum, as well as special mathematical topics (e.g., elementary algebra and calculus, cf. Aim 3c) and the analyses of extra-mathematical situations from the present or future professional fields of prevocational education and future fields of study for higher vocational education (cf. aim 1a). The most important topics in Berufsfachschule and Berufsaufbauschule are elementary algebra (for details see Bardy et al. 1988, pp.171-210) and - predominantly in technical colleges - geometry (pp.211-228). Fachoberschulen and Berufliche Gymnasien concentrate on calculus (pp. 229-254), analytic geometry and linear algebra (especially in technical domains) and statistics (for non-technical colleges). Often, these colleges have to treat junior secondary topics (like solving equations) extensively in order to fulfil the demands for manipulative competencies set by subsequent tertiary institutions (such as Fachhochschulen for future engineers or economists).

The dominant teaching method can be described as traditional solving of textbook problems. Knowledge of how to cope with certain classes of textbook problems (e.g., analysis of graphs of functions in calculus) is presented in a teacher-centred classroom interaction and then practised with numerous problems of the same type. Thus, teaching and learning are again centred on schemes and algorithms. Introducing concepts, argumentation and proofs, and epistemological reflection play only a minor role, if any. Realistic applications normally are not treated.

The reasons for this type of mathematics teaching in full-time vocational and technical colleges are partly the same as in part-time colleges: competencies ascribed to students as well as traditional types of examinations play an important role. Nevertheless, it is questionable whether this teaching is really appropriate for the students' careers and personalities. A broader scope of aims - especially by integrating more vocational and professional topics - and a different teaching style to build up appropriate fundamental ideas and the justified use of mathematical concepts in intra- and extra-mathematical contexts might be a better choice for teaching and learning mathematics in full-time vocational education.

3. MODELLING VERSUS LEGITIMATE PERIPHERAL PARTICIPATION

Most of the following arguments are taken from the Regular Lecture on 'Mathematics for Work - a Didactical Perspective' of Straesser at the 8th International Congress on Mathematics Education ICME 8 (cf. Straesser 1996).

3.1. Two pedagogies: modelling versus legitimate peripheral participation

If we take the organisational features of vocational education and training in Germany as an indicator of the underlying pedagogy, we find two extremes of learning principles: learning at the workplace versus classroom instruction (and in reality the standard oscillation and insecurity of political decisions on this matter).

Classroom type of vocational mathematics education tends to present mathematics as a separate body of knowledge, sometimes even structured along a disciplinary system from mathematics. In this case, mathematics has to be linked to work and workplace practice by building mathematical models and applying mathematics by the well-known modelling cycle of 'situation - (mathematical) model - interpretation of the situation'. The situation is to come from the workplace, the mathematical model rests upon mathematical structures and algorithms known before or taught on the spot and the solution of the model hopefully can be interpreted in a way to cope with the given professional situation (for a summary of this approach see Blum 1988). In this pedagogy, mathematics can come first and can be taught / learned along its own, disciplinary structure while applying it to work via modelling may come second, sometimes never or inappropriately. As can be seen from this description, the modelling approach clearly distinguishes two types of knowledge - namely professional and mathematical knowledge, which have to be brought together by the individual to cope with the professional problem. In most cases, modelling vocational problems by applying mathematics is a major difficulty for the future worker - especially the extraction of the mathematical model from a professional situation at hand.

The other extreme and contrasting pedagogy is training on the job, where learning takes place at the workplace whenever it is needed by the workplace practice and its problems. The focus is on coping with the situation at hand - and mathematics may come in or not when solving a workplace problem. An apprenticeship may offer a chance to legitimately participate in the workplace activities. At the beginning participation in work activities may be only peripheral (e.g. starting with minor preparatory or cleaning up activities). The duties of the apprentice will gradually shift from unimportant to activities essential for production or distribution. Without an explicit reference to the skills and knowledge necessary for coping with workplace requirements, without too formal or institutionalised teaching/learning, the new participant of the 'community of practice' gradually develops from a beginner to an expert at the workplace. With this approach, learning may be identified with taking part in the community of practice and gradually developing from a beginner to an experienced, full practitioner by means of situated learning (for a thorough discussion of the underlying concept of 'legitimate peripheral participation' see Lave & Wenger 1991). This pedagogy starts from a uniform concept of knowledge present in a community of practice (not in individual workers),

> knowing is inherent in the growth and transformation of identities and it is located in relations among practitioners, their practice, the artefacts of that practice and the social organisation and political economy of communities of practice (Lave&Wenger1991, p. 122).

As a consequence, mathematics can continue to go invisibly, embedded in the workplace practice and serving as a tool used to cope with professional problems if needed. A problem-oriented integration of concepts tends to hide mathematical relations under the uniform workplace practices. Following this approach, studies on 'street mathematics' (like Nunes et al. 1993) had to detect and bring back to light the mathematical procedures in workplace activities, to describe them and to show the competence of the practitioners in using mathematics.

In some sense, the 'Rich Interpretation of Using Mathematical Ideas and Techniques (RIUMIT)' project takes a somewhat intermediate approach: The teacher-researchers 'became conscious of the dangers of taking the mathematics out of context' (a denial of a pure modelling approach) and the project found that 'neither mathematics nor the skills involved in using the mathematics occur in isolation from the situation, other knowledge and other generic skills' (a statement very near to the legitimate peripheral participation approach, both quotations from AAMT 1997, p. 21f). Nevertheless, the same document does not opt for a disappearance of mathematics from the curriculum, but tries to foster a change within the given curricula and/or those under development (loc.cit., p. 68ff).

3.2. *An aside on 'transfer'*

The starting point of the research on 'street mathematics' was a twofold observation: (a) With little or no schooling, the children working in the streets were able to solve their 'mathematical' problems at work. (b) Even if the children had attended school, children successful at work could not or only with difficulties solve 'isomorphic' college type word problems. How to understand this obvious lack of transfer from classroom to work ?

After more than a decade of research (cf. the monographs Lave 1988, Lave & Wenger 1991, Nunes et al. 1993, Saxe 1991), the protagonists of situated learning in a community of practice can easily understand the dilemma described above: Mathematics used in the street is learned there, is efficient in solving the street problems and fundamentally different from the one learned in school or researched in a mathematics department at university. To rephrase it in a more general way: Mathematics learned in a specific context is part of a subjective domain of experience (cf. Bauersfeld 1983) and cannot easily be isolated, taken away, transferred and applied in a different situation.

In contrast to that, the 'mathematical modeller' starts from the assumption that a piece of mathematics once learned will come to mind whenever it models (adequately) a given situation, that - after appropriately modelling the situation - it can be applied easily and will offer a decent way to cope with the problem at hand. In fact, reality seems to be less convenient: The learner usually has difficulties to mobilise her/his knowledge in so-called isomorphic situations, s/he has problems to transfer a procedure, a solution from one situation to a different, maybe unknown, one. Eberhard (in this book) adds an analysis of problems coming from a bad distribution of teaching certain pieces of knowledge between workplace and classroom. The widespread and well documented lack of easy transfer definitely contradicts the plea for modelling and easy application.

3.3. *Proposing a way forward*

As a consequence of a preference for the situated learning and community of practice approach, why not dissolve any classroom type of training at least in vocational mathematics and totally rely on training on the job for vocational mathematics? I want to draw your attention to a finding which might be forgotten when closing vocational / technical colleges: In the study on Brazilian bookies, the protagonists of street mathematics state :

> [...] the influence of schooling is not limited to topics explicitly taught in classrooms but [...] school experience provides a different way of analysing and understanding everyday activities. [...] Schooled bookies [...] seem to have a different attitude toward procedures for solving problems as a result of their schooling. [...] school experience has an effect on how people deal with more academic questions, such as explaining their everyday procedures or making explicit the mathematical structures implicit in their everyday activities. School experience is also related to better performance on solving problems that differ from those usually encountered at work (Schliemann&Acioly 1989, p. 216 ff.).

Obviously, classroom type of activities can offer an opportunity to broaden the perspective of the future worker, to empower her/him with solving problems not common to workplace practice and to foster understanding of the workplace procedure. Classroom type of activities can offer an understanding which goes beyond the narrow confines of the actual situation, which transcends the situation and the problem where and when knowledge is developed. Classroom type activities in schools or colleges can show mathematics as a way to transcend the context with more general problem solving strategies and structures. But how to cope with the transfer dilemma described in part 3.2?

As far as I can see there is a 'way out'. Modelling with the help of mathematics should not be taken as a means to get rid of the dirty specialities of the concrete workplace to solve the abstracted problem by means of pure mathematics. It is by exploiting the interplay of the professional, concrete situation and the structural, mathematical model that one can cope with the given professional problem. In doing so, one can develop a mathematical structure maybe adaptable to a variety of different problems linked to the initial professional situation. (to set up a 'domain of abstraction' where the 'dialectic between concrete and abstract' closely ties together mathematical ideas and practical knowledge of the professional domain, cf. Noss&Hoyles 1996., p. 27; see also Noss et al. in this book). In doing so, mathematics is not reduced to the general type of activity of theorising, analysing language and seeing structures implicitly devaluing situated learning as learning no mathematics (for a recent claim of this reduced point of view cf. Sierpinska 1995, p. 5).

If mathematics is taught as a bridge between the concrete, may-be vocational situation and the abstract, may-be systematic structure, even classroom vocational education can show mathematics as a 'general' tool which is of larger an importance than just coping with the narrow tasks of the everyday work practice or the inculcation of algorithms. If college type education aims at presenting (vocational) mathematics in this way, one condition for success seems to be that mathematics is taught in a way it is 'meaningful to the individual' who is learning. Technical and vocational colleges then have to strive for problems from the workplace which are as realistic as possible. And the problems should be taught in a way as close as possible to the actual concerns of the students (for an elaboration of this cf. Boaler 1993 and Evans in this book).

An additional case for learning mathematics not in a too narrow workplace context is expressed in a reminder I would like to place at the end of this section:

> Mathematics in vocational education is serving more as a background knowledge for explaining and avoiding mistakes, recognising safety risks, judicious measurement and various forms of estimation. [...] Not practice at the workplace but deepening of the professional knowledge, education to a responsible use of tools and machines and the understanding of and coping with everyday mathematical problems legitimise mathematics in vocational education' (Appelrath 1985, p. 133/139; translation R.S.).

4. MATHEMATICAL AND OTHER MEANS IN VOCATIONAL CONTEXTS

In parts 2 and 3 of this paper, we find numerous hints to phenomena reported by nearly all researchers on mathematics in the workplace (e.g. Bromme et al. 1996 or the Cockcroft-report 1982):

-workers often deny to use any mathematics at all at the workplace (apart from simple arithmetic),

-mathematics is condensed in algorithms and routines which are used without noticing that they are of mathematical nature,

-mathematics seems to gradually disappear from the workplace.

How to understand this development and how to cope with it in (general and vocational) education and training ?

4.1. Means ('primary' artefacts)

A recent study on workplace use of mathematics came up with the following finding:

> Practitioners at work employ elements of mathematical knowledge in a fragmented way as strategies are finely tuned to particular tasks. These strategies frequently involve the use of artefacts which in some sense are already mathematised - indeed, their functionality depends on a lack of awareness of the mathematical model underlying them. (cf. Noss et al. 1998, p. 14; for the same finding see for instance AAMT 97, p. 55; on artefacts in general cf. Wartofsky 1979, especially p. 201ff).

In my view, the use of artefacts is a most important aspect of mathematics at the workplace. This aspect is often overlooked, but offers a perspective to explain the above mentioned phenomena:

If the job runs smoothly and routinely without unfamiliar and unforeseen events, practitioners tend to rely on well-known routines for repetitive problems. These routines are often implemented in tools (like machines for calculating, scales to read, charts to fill etc., i.e. 'primary' artefacts sensu Wartofsky). Difficulties when using mathematics tend to be simplified, if not totally avoided by algorithms and routine flows of activities. Bookkeeping with its since long formalised set of concepts and practices (like discount and increase, recording of transactions by means of accounts, book-keeping by double entry etc.) can serve as an illustration how complicated workplace practices are routinised by 'simple' algorithms. As long as the workplace does not present unexpected situations, these artefacts go unrecognised and 'hide' the mathematics they incorporate.

The study by Noss et al. 1998 additionally presents findings on non-routine situations - either deliberately produced by the researchers or occurring during 'normal' working hours. In such situations of 'breakdowns' or unfamiliar settings, practitioners 'tend to [...] apply a fragment of professional knowledge, a half-remembered rule from school mathematics or a novel, though generally unsuccessful, use of a familiar tool' (Noss et al. 1998, p. 14; also Noss et al. in this book; for the non-understanding of the workplace mathematics cf. also Hogan 1996, p. 288). Here again, the artefacts show up as one way to somehow manage non-routine problems - and the 'banking mathematics study' shows that computers and finely tuned software can even be used to offer a microworld for exploring non-routine, unusual situations (cf. Noss&Hoyles 1996; for a detailed discussion of computer technology cf. part 4.3 of this paper).

If we take 'artefacts" in the broad sense of Wartofsky, we can describe the trend of hiding mathematics in algorithms and routines in an easy way: By integrating mathematical concepts, relations and procedures into various types of primary artefacts (be it rules to follow, charts to fill in, computer technology to handle or other machines to use) mathematics tends to gradually disappear from the attention of the worker. Even if mathematics is increasingly used on a global level, the individual professional does not notice this development, s/he tends to describe the ongoing process as a gradual disappearance of mathematics from her/his workplace. What about teaching/learning mathematics under these circumstances?

4.2. Teaching and means ('secondary' artefacts)

If we take into account the utmost importance of artefacts in vocational contexts and the hiding of mathematics behind or 'inside' primary artefacts, teaching vocational mathematics becomes difficult. In some sense, we run into the same dilemma as the one in part 3 of this paper: Either the teaching process isolates and extracts the mathematics incorporated in a workplace procedure, in a 'primary' artefact. As a consequence of this process, the unity of the workplace knowledge is destroyed, the 'gestalt' of the workplace activity is decomposed into various, distinctive parts, vocational mathematics has lost its 'authenticity'. If taught like this, the learner will have difficulties in recognising the mathematics taught as the vocational mathematics used in a vocational tool. The modelling approach in part 3.1 is one way to retain the vocational context and make mathematics visible at the same time. A different approach to teaching would be preserving the unity of the workplace activity and the artefacts used at work. Such a presentation of the vocational tool as such without explicitly looking into or explaining (and showing the general applicability of) the mathematical aspects of the tool is obviously nearer to an legitimate-peripheral-participation approach. It has the advance of being more authentic - but hides the scope of applicability and the generality of the mathematics used.

If we look into a classroom type of teaching, the situation is even more complicated. In addition to the problems of finding and learning mathematics at the workplace, the classroom changes the community of practice from the team at the workplace to the peers of the classroom. Even more important, the artefacts which are handled in the classroom are (usually) not the same as those applied at the workplace. To cope with this change, Wartofsky introduces 'secondary' artefacts as 'those used in the preservation and transmission of the acquired skills or modes of action or praxis [...] Secondary artefacts are [...] representations of such modes of action' (loc.cit., p. 202) and by this very nature different from the modes of action or skills itself. Compared to the primary artefacts, the secondary artefacts used in the teaching process have undergone a 'didactical transposition' which implies certain changes to the knowledge used at the workplace, namely a compartmentalisation, depersonalisation, decontextualisation and sequentialisation of the knowledge taught in comparison to the knowledge used in the workplace (cf. Chevallard 1991, chapter 5, the remark on 'differences between experiences of learning in the work place and in educational settings' in Hogan 1996, p. 288, points into the same direction). Naturally, this 'didactical transposition' also alters the mathematics incorporated in workplace procedures and modes of action and changes the nature of the mathematics taught in classrooms as compared to mathematics used in the workplace. The didactical transposition necessarily creates a distance between what is actually learned in a classroom and the primary artefacts of the workplace. This may result in students not knowing about the workplace practice (and primary artefacts) in a useful way - but is inevitably linked to the difference between the workplace and the classroom teaching.

4.3. Teaching and learning with information technology

Obviously, a very versatile and widely useful artefact can be modern information technology: computers, appropriate software and other electronic devices. Nevertheless, this technology can have ambivalent effects in the workplace and in vocational training and education.

Using finely tuned information technology can hide the mathematics incorporated in the software. Noss&Hoyles 1996 precisely describe the effects of banking software which has been developed by highfly specialists and completely hides the inherent functional, mathematical relations between the numbers processed. Even in the number driven world of banking, numbers and commercial arithmetic disappear from the consciousness of the average employee. Mathematics hide in computer algorithms which are applied without paying attention to the underlying mathematical model of the banking process. Even somewhat complicated procedures (like calculating the present value of a treasury bill by discounting from face value in dependence of the day of maturity) go unrecognised by the average employee who relies on the programs designed by an unknown specialist in an unknown software house or department. The concepts of the users at the workplace reduce these numbers to mere indices of banking information, the underlying mathematics is totally blended out and ignored.

> [...] these models were almost entirely hidden from view. Understanding and reshaping them was the preserve of the rocket scientists; the separation between use and understanding was absolute and the models' structures were obscured by the data-driven view encouraged by the computer screens' (cf. Noss&Hoyles 1996, p. 17).

Additional examples could be the multitude of functional and numerical relations which usually go unrecognised when a spreadsheet is used. Most computer based accounting systems act in the same way and do not ask for a deep understanding of the double entry approach. CAD software may automatically change the perspective drawing of a 3D-object or a virtual reality by trial-and-error control without even asking for an understanding of the underlying concepts like axial versus central perspective and / or different types of modelling 3D-objects. In most workplaces, the use of modern (computer) technology implies the use of sophisticated mathematical models - but these models go without recognition by the average employee.

On the other hand, the paper by Noss&Hoyles shows, that exactly the same technology can be used to de-grey the black boxes, to show the mathematical relations and offer an opportunity to explore the inherent, implemented relations in a way, workplace reality would never allow because of the risk of material, financial and time losses. The practice of using sophisticated mathematics can be brought to the foreground and consciousness of the user by appropriate courses. And it is modern computer technology and appropriate software again which can be successfully used in this process to really explore and understand the underlying banking mathematics.

To sum up: Modern computer technology itself has a dual role in the process of using mathematics at the workplace: It can be used as a way to hide mathematics in sophisticated software. Mathematics as a tool, a man-made artefact disappears in workplace routines - and modern technology can speed up this disappearance. On the other hand, the very same technology can be used to foster understanding of the professional use of mathematics by explicitly modelling the hidden mathematical relations and offering software tools to explore and better understand the underlying mathematical models.

5. CONCLUSION

5.1. Lessons to learn for general mathematics education

From a picture like this, there may be lessons to learn for general, non-vocational mathematics education:

-It is (very) difficult to identify mathematics as used in the workplace. Even if it is a commonplace that mathematics is becoming more and more important because of its vocational and societal use, it can be difficult for a (mathematics) educator to 'show' the growing use of mathematics to students, parents, curriculum developers and legislators.

-It is difficult, if not impossible, to teach mathematics in the classroom in the same way as it is used in the workplace. There are two barriers to this: (1) Teaching in the classroom has constraints different from the constraints of the workplace. It has to transpose the knowledge to be taught into a decontextualised, depersonalised and sequential knowledge in order to make it teachable. Teaching mathematics uses 'school-mathematics', i.e. 'secondary' artefacts as compared to 'primary' artefacts at the workplace. (2) Teaching mathematics aims at presenting a true picture of mathematics - as a societal tool to cope with extra-mathematical problems and as a scientific discipline. Aiming at new mathematical knowledge and the best inner, logical structure of knowledge, mathematics as a scientific discipline totally changes and reorganises mathematics as it is used in the workplace.

-Learning at the workplace tends to follow the legitimate peripheral participation approach in a community of practice. Learning mathematics as a scientific discipline seems to follow different lines with longer phases of splendid isolation for the individual learner (cf. Sfard et al. 1998). To my knowledge, there is no detailed research on this difference and on the consequences for teaching/learning mathematics up to now.

-There is an urgent need to describe and defend the authenticity of the classroom against different authenticities from other sources (be it a scientific discipline or the variety of workplaces). New technology with the opportunities of exploration via simulation can help with looking outside the classroom without imitating mathematics as a scientific discipline, mathematics in the workplace or in society at large.

5.2. *Speculating on the future*

During a seminar for New South Wales TAFE-teachers on September 6, 1984, I put forward trends 'in mathematics education for vocational use' (cf. Straesser 1985, p. 13f). One of these speculations was 'Learning by doing becomes more difficult! Using mathematics becomes more important!' About 15 years later I would like to comment and elaborate this look into the crystal ball:

If I start with the growing importance of mathematics, my first guess still is that mathematics is getting more and more widely used and getting more important - BUT: the growing importance of mathematics is paralleled by a growing integration of mathematics into a variety of non-mathematical artefacts. As a consequence of this integration (better: disappearance?) of mathematics in more and more sophisticated tools, it can be difficult to advance or defend teaching and learning mathematics by referring to its (growing) importance. The use of high technology tools can often be taught/learnt via showing-how-to-do and imitation at the workplace and does not presuppose understanding. Only if we as educators and citizens stick to the vision of a skilled worker and informed citizen who at least understands the basics of her/his workplace and society we can also opt for the teaching and learning of more mathematics to look behind the surface of highly integrated and sophisticated tools and modes of social organisation.

The growing difficulty of learning by doing is true only in highly automated, technologically advanced areas of production and distribution. If available, information technology may even open a chance to learning by doing in these advanced types of vocational contexts - if we use a wider idea of 'doing' and if we intelligently use the opportunities of virtual reality and simulation. The rise of legitimate peripheral participation as a new and different model of teaching/learning to me is an additional hint that my speculation on learning by doing may no longer be true.

Acknowledgement
The text is a slightly revised version of a paper first published in the conference proceedings Baldwin, K. & Roberts, J. (eds.): Mathematics - the next milleneum. Adelaide (Australian Association of Mathematics Teachers AAMT) 1999, p. 17-29. Reproduction with kind permission from AAMT.

6. REFERENCES

Appelrath, K.-H. (1985). Zur Verwendung von Mathematik und zur Situation des Fachrechnens im Berufsfeld Metalltechnik (dargestellt an zwei Unterrichtsbeispielen). In P. Bardy, W. Blum, H.-G. Braun, (Ed.), *Mathematik in der Berufsschule - Analysen und Vorschläge zum Fachrechenunterricht* (pp. 172-139). Essen: Girardet.

Australian Association of Mathematics Teachers (1997). *Rich Interpretation Using Mathematical Ideas and Techniques (RIUMIT) - Final Report.* Adelaide: Australian Association of Mathematics Teachers.

Bardy, P., Blum, W., and Braun, H.G. (1985). *Mathematik in der Berufsschule - Analysen und Vorschläge zum Fachrechenunterricht.* Essen: Girardet.

Bardy, P., Blum, W., Market, D., and Strässer, R. (1988). Mathematik in der Beruflichen Bildung. In P. Bardy, et al, (Eds.), *Technic Didact Vol. 3* (pp. 141-254). Alsbach: Leuchtturm (Diskussionsfeld Technische Ausbildung).

Bauersfeld, H. (1983). Subjektive Erfahrungsbereiche als Grundlage einer Interaktionstheorie des Mathematiklernens und-lehrens. In H. Bauersfeld, et al (Eds.), *Lernen und Lehren von Mathematik* (pp. 1-56). Köln: Aulis-Verlag Deubner.

Blum, W. (1988). *Theme Group 6: Mathematics and Other Subjects*. Paper presented at the Sixth International Congress on Mathematical Education, Budapest.

Blum, W., and Straesser, R. (1992). Mathematics Teaching in Technical and Vocational Colleges - Professional Training versus General Education. In H. Schupp (Ed.), *Mathematics Education in Germany* (pp. 242-247). Karlsruhe: ZDM.

Boaler, J. (1994). The Role of Contexts in the Mathematics Classroom: Do they make Mathematics More "Real"? *For the Learning of Mathematics, 13*(2), 12-17.

Bromme, R., Rambow, R., and Straesser, R. (1996). Jenseits von 'Oberflache' und 'Tiefe': Zum Zusammenhang von Problemkategorisierungen und arbeitskontext bei Fachleuten des Technischen Zeichnens. In H. Gruber and A. Ziegler, (Eds.), *Expertiseforschung - Theoretische und Methodische Grundlagen* (pp. 150-168). Wiesbaden: Westdeutscher Verlag.

Bundesministerium für Bildung, Forschung und Technologie. (1997). *Grund- und Strukturdaten Ausgabe 1997/98*. Magdeburg: Gebr. Garloff GmbH Druckerei und Verlag.

Chevallard, Y. (1991). *La Transposition Didactique - du Savoir Savant au Savoir Enseignè, (2nd edition)*, Grenoble: La Pensée Sauvage.

Cockcroft, W. H. (1982). *Mathematics Counts: Report of the Committee of Enquiry*. London: HMSO.

Hogan, J. (1996). *A Methodology for Using Classroom Teachers to Gather Information about Using Mathematics from the Work Place*. Proceedings of the 19th Annual Conference of the Mathematics Education Research Group of Australasia (Merga), Melbourne.

Lave, J. (1988). *Cognition in Practice: Mind, Mathematics and Culture in Everyday Life*. Cambridge: Cambridge University Press.

Lave, J. and Wenger, E. (1991). *Situated Learning: Legitimate Peripheral Participation*. Cambridge: Cambridge University Press.

Noss, R., and Hoyles, C. (1996b). The Visibility of Meanings:, Modelling the Mathematics of Banking. *International Journal of Computers for Mathematical Learning, 1*(1), 3-31.

Noss, R., Hoyles, C., and Pozzi, S. (1998). *Towards a Mathematical Orientation through Computational Modelling Project*. ESRC End of Award Report: Mathematical Sciences Group, Institute of Education, University of London.

Nunes, T., Schliemann, A., and Carraher, D. (1993). *Street Mathematics and School Mathematics*. Cambridge: Cambridge University Press.

Saxe, G. (1991). *Culture and Cognitive Development: Studies in Mathematical Understanding*. Hillsdale NJ: Lawrence Erlbaum Associates.

Schliemann, A. D., and Acioly, N. M. (1989). Mathematical Knowledge Developed at Work: The Contribution of Practice versus the Contribution of Schooling. *Cognition and Instruction, 6*(3), 185-221.

Sfard, A. P., Nesher, P., Streefland, L., Cobb, P., and Mason, J. (1998). Learning Mathematics through Conversation: Is It as Good as They Say? *For The Learning of Mathematics, 18*(1), 41-51.

Sierpinska, A. (1995). Mathematics: "in Context", "Pure", or "with Applications"? A Contribution to the Question of Transfer in the Learning of Mathematics. *For The Learning of Mathematics, 15*(1), 2-15.

Straesser, R. (1982). *Mathematischer Unterricht in Berufsschulen - Analysen und Daten*. Bielefeld:Universität Bielefeld: IDM, Materialien und Studien.

Straesser, R. (1985). *Mathematics in Technical and Vocational Education in the Federal Republic of Germany (FRG) Occasional Paper Nr. 62)*.: Institut für Didaktik der Mathematik (IDM) der Universitat Bielefeld.

Straesser, R., Barr, G., Evans, J., and Wolf, A. (1989). Skills Versus Understanding. *Zentralblatt für Didaktik der Mathematik, 21*(6), 197-202.

Straesser, R. (1996). *Mathematics for Work - a Didactical Perspective (Regular lecture)*. Paper presented at the ICME 8, Sevilla.

Sträesser, R., and Zevenbergen, R. (1996). Further Mathematics Education. In A. Bishop, K. Clements, C. Keitel, J. Kilpatrick and C. Laborde, (Eds.), *Mathematics International Handbook on Mathematics Education* (pp. 647-674). Dodrecht: Kluwer.

Wartofsky, M. W. (1979). *Models, Representation and Scientific Understanding*. Dordrecht: Reidel.

Wolff, F. W. (1958). Ziel, Stoff und Weg im Fachrechnen der Gewerblichen Berufsschule. In F. Drenckhahn (Ed.), *Der Mathematische Unterriicht für die Sechs- bis Fünfzehn- Jährige Jugend in der Bendesrepublik Deutschland* (pp. 195-200). Göttingen: Vendenhoek and Reprecht.

SECTION 2

BRINGING SCHOOL AND WORKPLACE TOGETHER

SECTION 2

BRINGING SCHOOL AND WORKPLACE TOGETHER

PREFACE

SUSAN L. FORMAN AND LYNN ARTHUR STEEN

Historically, the relation between school and workplace has rested on paradox. Whereas many adults expect schools to prepare students for work, most of those who teach in schools have never worked anywhere but in school. So teachers teach what they know—which is how to prepare for more school, not for the real world of work. Moreover, parents who desire that their children leave school well prepared to earn a living in the modern economy nonetheless expect schools to cover pretty much the same curriculum as they did when they themselves were students.

In mathematics the paradox is compounded by the isolation of advanced school mathematics from both public discourse and workplace expectations. Even as educators, parents, and politicians exhort pupils to learn ever-increasing amounts of mathematics, many studies show that adults rarely use (and certainly don't remember) much of the mathematics they learned in secondary school.

In this section these paradoxes are explored from four very different perspectives in different national contexts - the Netherlands, Australia, the United States, and Denmark. Two fundamental questions are addressed in each of these papers:

-What mathematics meets the needs of workers?

-How can this mathematics be taught in school?

In the first paper, Pieter van der Zwaart reports on the SLO project in the Netherlands that seeks to determine how mathematics can contribute to slow learners' preparation for future jobs. By gathering information from craftsmen and experts in various trades, this project revealed two little-recognized but widespread characteristics of mathematics in the workplace. First, there is more mathematics in the workplace than people expect or readily recognize, even in jobs performed by students who have had to struggle to learn mathematics. Second, the narrow experience of school mathematics has so limited the mathematical understanding of many adults that even experienced workers often fail to recognize the mathematics they actually employ.

Bessot & Ridgway (eds.), Education for Mathematics in the Workplace, 83—86.

The SLO researchers discovered that workplace mathematics falls naturally into four categories: three-dimensional thinking, use of complicated information systems, measuring and calculating diverse amounts, and following rules and procedures. Typically, only elementary and non-formalized mathematics is involved in each of these categories, but the situations in which the mathematics is used are usually fairly complex.

To test their hypotheses about teaching mathematics for the workplace, the SLO staff translated several workplace scenarios into 'story-lines' for teachers to use in the classroom. These proved quite effective when motivated by well-prepared teachers, but less so when students were expected to learn from the teaching materials alone. Student work on projects based on these work-based story lines was serious and professional, in contrast to slow learners' typical attitude towards traditional school mathematics.

John Hogan and Will Morony describe a somewhat similar effort in Australia, but with important differences. One of several 'key competencies' defined in a 1992 Australian report on employment and training is 'using mathematical ideas and techniques'. Because people often ignored or misinterpreted this competency - probably due to the widespread adult avoidance of mathematics - the Australian Association of Mathematics Teachers launched a project called 'Rich Interpretation of Using Mathematical Ideas and Techniques' (RIUMIT). The goal of this project was to generate descriptions ('stories') of how mathematics manifests itself in workplace situations.

The RIUMIT project was carried out by mathematics teachers who engaged in highly structured job-shadowing followed by interviews that led to written descriptions of workplace tasks that highlighted the embedded roles of mathematics. The goal of these teachers-turned-researchers was to describe what happens in work situations without imposing any predetermined view of mathematics. Hogan and Morony's paper describes the protocol of job shadowing in the RIUMIT project, which itself turned into a marvelous professional development program for the teachers.

The RIUMIT project uncovered many of the same attitudes about mathematics that were reported by the project SLO researchers. For example, when business leaders were asked to arrange job-shadowing placements, they invariably selected mathematics-based jobs in accounting or management because they didn't recognize mathematics as being used in any other work in their business. However, when teachers shadowed other workers, they discovered a lot of mathematics being used that the worker neither recognized as mathematics nor connected in any way with the mathematics learned at school. (Some of the thirty stories produced by the RIUMIT researchers can be found on the Internet at www.aamt.edu.)

In the third paper, Forman and Steen describe contrasting (and sometimes conflicting) attempts in the United States to improve both the occupational skills of entry-level employees and the mathematics education for all students, many of whom now drop out of mathematics as soon as it is no longer required. Both initiatives employ the mechanism of 'standards', although the meanings of this term, as well as its implementations, differ significantly among various occupational and academic areas.

The standards for school mathematics developed by the National Council of Teachers of Mathematics (NCTM) recommend a substantial core of mathematics for all students, including increased emphasis on statistics and data analysis. In contrast, most occupational skill standards describe more limited mathematical expectations. Yet both sets of standards agree on their vision of effective teaching and learning. Both urge that mathematics be embedded in rich, authentic contexts and that students experience, explore, and explain the mathematics they encounter.

Several tensions are exposed by the differences in outlook between leaders of the occupational skills movement and the NCTM - led mathematics educators. Foremost is the issue of equity and tracking, an issue of special significance in a multicultural nation like the United States. Second is the matter of curriculum and pedagogy- whether it is more or less likely that students will learn the mathematics they need for higher education if they focus on authentic work-based tasks rather than traditional decontextualized 'word problems'.

In the Netherlands, Australia, and the United States, as in most nations, most people assume that mathematical preparation is essential for jobs. Thus projects such as SLO, RIUMIT and the various U.S. standards efforts do not ask whether mathematics is indeed necessary, but rather seek to clarify the kinds of mathematics required and then to implement these kinds of mathematics in curricula. In the final paper of this section, Tine Wedege of Denmark examines the theoretical underpinnings of the relation of mathematical knowledge to workplace skills. As Wedege argues, mathematical knowledge does not become a true vocational qualification unless it is integrated with other factors that are essential in the workplace.

Wedege distinguishes three kinds of qualifications for employment: general (literacy and numeracy), professional (technical skills), and social (personal characteristics). Mathematics is intimately involved with the first two types of qualifications. Depending on the workplace context, the same kind of mathematics may sometimes be a general qualification and at other times a professional qualification.

The dual nature of mathematical knowledge is reflected in Wedege's analysis—mathematics as a discipline of knowledge, and mathematics as one of many components of a job situation. These differ significantly in the educational contexts in which they are taught - the former being the focus of traditional academic education, the latter of apprenticeship or vocational education. The situation in Denmark, although perhaps more systematic and pervasive, is not unlike that in many other countries. In addition, Wedege argues, the use of mathematics in the workplace is also subject to the (judicious) influence of attitudes, feelings, experiences, and motives in deciding how to interpret and act on information.

Despite differences in national context and research purposes, these four papers convey several common characteristics about workplace mathematics that are of crucial importance to anyone seeking better connections between school and workplace:

-Current literature is almost devoid of good examples of secondary school mathematics used in authentic workplace settings.

-Mathematics teachers have virtually no experience using mathematics in real work contexts.

-Workers' experience with mathematics in school limits their ability to perceive mathematics used at work.

-Whereas context is generally irrelevant to school mathematics, in the workplace mathematics is inextricably embedded in context.

-Workplace tasks typically require sophisticated problem solving using primarily elementary mathematics.

For these reasons and others, mathematics at school is quite unlike mathematics at work. The task of reconciling these two enterprises requires considerable effort and much research. Among the many issues that need further exploration are the following:

1. How can the mathematics used at work be identified, described, and conveyed to teachers?
2. How can authentic, contextualized tasks be deployed in widely-adopted curricula when their authenticity depends on local details and situations?
3. Do students learn mathematics better from rich tasks that reflect real work situations or from more didactic lessons that focus on the mathematics itself?
4. Can a single curriculum serve all students equally well--those who expect to attend university as well as those who will enter the workforce directly?

Further work on questions such as these should help teachers motivate students to learn more mathematics in a usable form, thus lessening the gulf between school and workplace.

CHAPTER 6

WORKING MATHEMATICS FOR LEARNERS WITH LOWER ABILITIES

AN INVESTIGATION OF HOW TO INTEGRATE MATHEMATICS FOUND ON THE FUTURE SHOP FLOOR, INTO THE MATHEMATICS LESSONS OF LEARNERS WITH LOWER ABILITIES IN SECONDARY EDUCATION

PIETER VAN DER ZWAART

Abstract. What mathematics is of importance to learners with lower abilities from the point of view of their future profession? This article describes the approach of the project-group on mathematics in the National Institute for Curriculum Development (SLO) in its search to find answers to this question. The strategy selected and, within this, the role of professionals, the functioning of experimental teaching materials and analyses will be discussed thoroughly. The effects of the outcome of this project on official papers, such as curricula and examination syllabi, will be considered as well.

1. INTRODUCTION

The May 1989 issue of the 'Nieuwe Wiskrant', a Dutch magazine on mathematics education, included an article by José de Haan (Haan, J. de, 1989)in which she describes, very inspiringly, how she and her students work during math classes. The students are all girls and education concentrates on lower domestic sciences. Apparently, the professional perspective of these students plays a major role during their math classes as José de Haan, on behalf of her students, finishes her contribution with a cry of distress, meant for all who are concerned with changes in the education of mathematics: 'What do I need math for, if I want to be a hairdresser, maternity welfare worker, etc......??????'

In 1993 in the Netherlands, major changes in the mathematics curriculum and examination syllabi for secondary education took shape. The former curriculum dating from 1968, included some clear "new math" characteristics. It was replaced by a newly developed curriculum and examination syllabus that can be characterized as real world math (See for instance: Lange, 1996).

However, with respect to learners with lower abilities one question remained unanswered: What part does mathematics education play in the preparation for future trades? In particular in the case of learners with lower abilities. They are

Bessot & Ridgway (eds.), Education for Mathematics in the Workplace, 87—100.

expected to find, within the apprentice system, jobs and set foot on the shop floor once they have left secondary education.

The above describes two critical moments in the discussion on the role of mathematics in secondary education in the preparation of learners with lower abilities for their future jobs, that has been going on in the Netherlands for some time now. These moments are about drafting the problems at the level of the pupil and at that of the curriculum.

As a result of changes in the curriculum and in consultation with the Dutch association of Mathematics Teachers SLO started a project in 1992 with the following question as the major issue: How can mathematics education contribute to the preparation for the future profession? The answer to the question should apply to students in the third and fourth grades of pre-vocational education[1].

2. DESIGNING OF THE INVESTIGATION[2]

One of the first questions was: How do we approach the relationship between mathematics and occupation? The following points of view were taken into consideration:

-Consider the routine activities of a certain trade as a source of inspiration for mathematical activities.

-Find out which mathematical topics are essential in the training for a certain profession.

-Find out what mathematical ingredients there are in the problems experienced on the shop floor.

The first point of view mentioned is prominent when the development of mathematical knowledge and skills themselves are considered most important. However, it is the flexible use of previously acquired mathematical knowledge and skills that is of importance during the third and fourth grades of pre-vocational education. Also starting from this point of view mathematical subjects may occur that are not in use on the shop floor[3].

For instance, in cases requiring perpendicular lines and planes, the introduction of Pythagoras seems obvious. Thus you may find yourself mathematical inspired on a building site. However closer investigation makes clear that in practice not one single construction worker avails himself to Pythagoras' theme. Construction workers know they can lay off a right angle with the help of a triangle with sides 3, 4 and 5. For checking perpendicularity they know the diagonals of a rectangle are equally long. So it is hard for a mathematics teacher to answer the question: 'What use is Pythagoras' theme if I want to be a construction worker?' Clearly this is not a plea to drop Pythagoras. Getting inspired by the building site is a very legal argument to perform mathematical activities. However, legalising Pythagoras by presenting a construction worker using it is far beside the truth. Unfortunately it is often seen in teachingmaterials that authors walk straight into this trap.

At first sight the second point of view seems a good one to start with. However, a pilotstudy soon made clear that mathematical education in vocational training is strongly affected by the traditional approach of mathematics education. Furthermore many examples like the one described above were found. Only very little of the mathematics offered there, is actually found on the shop floor.

Eventually it was the third approach that was opted for. The next section describes the exploratory study that was carried out and which was also the basis for the option that was made. There were various reasons:

It might well be that mathematics straight from the shop floor meets the abilities of this group of learners and that it has motivating effects.

Pilot studies made clear that the future shop floor of learners with lower abilities is rich in activities with mathematical aspects. The subject has hardly been described, though.

Furthermore, the national core curriculum (Miow, 1994) stated:

> In relevant situations, the pupils can see the connection with the practice of some professions.....They obtain some insight into the meaning of each subject in subsequent studies and professions.

The research question was formulated as follows: Is it possible to introduce shop floor mathematics in the learners mathematics programme for learners with lower abilities? This question was answered at two levels.

Model lessons demonstrating shop floor mathematics were developed and tried out in classroom practice. First of all this was done to show that the successful trying-out of experimental teaching materials supports the feasibility of the underlying ideas at the level of the mathematics class. The other reason is that these materials and the experiences that come with them make an essential contribution to the process of formulating the intentions of and the discussions on the underlying assumptions.

An analysis (Zwaart, 1997b) comparing the contents of the standard curriculum and that for the future shop floor of learners with lower abilities. On the one hand the purpose of this analysis was to establish the knowledge gained in this project at a theoretical level. On the other hand the analysis was meant to support the discussion on the selection of contents for curricula and examination syllabi.

Thus, the research question, in fact, turned into a designer problem: Is it possible to develop teaching materials meeting the research question? If so, then together with a description of conditions essential for successful realization, the research question has been answered.

Career opportunities for learners with lower abilities are vast and greatly varying. This also holds for the numerous course programmes teaching them to carry out specific jobs. The full mapping of opportunities and programmes with respect to the uses of mathematics was beyond our scope. Therefore, choices had to be made.

It was decided to select a theme requiring the participation of three occupational groups and then determine the mathematics relevant to the theme in each of the groups. Having determined the relevance of certain mathematical subjects and skills, the next step would be the development of experimental teaching materials. Should this set-up turn out to be one leading to positive results, in order to validate our approach another theme would have to be selected requiring three more occupational groups.

The first theme selected was the installation of a kitchen in an existing home. There are obvious parts for construction and installation work and for salesmanship (Hove, 1996). The other theme, worked out at a later stage, concerns the infrastructure of a large hospital. From the many sectors possible we selected mechanics, food and healthcare (Zwaart, 1996). Major criteria determining the process of selection were:

All combinations of occupational sectors should contain interesting aspects for pupils from different groups: boys\girls, indegenous\immigrants.

A sector should not be too specialized, neither in a sense that only few opt for work in that specific sector, nor that the mathematics there hardly occurs on any other shop floor.

Variety in the number of mathematical subjects as well as regular applications, both relevant and learnable for learners with lower abilities.

3. EXPLORING THE SHOP FLOOR

Information concerning the use of mathematics on the shop floor was gained from craftsmen, foremen and superiors in different trades. Without exception all, of them showed great kindness and were ready to tell about their trades and about the day-to-day practice of the work. However, this was only part of the solution to the problem of gathering the right information. It turned out that the picture tradespeople had of mathematics was often highly coloured by their experiences at school. In consequence they often could hardly recognize the mathematical aspects of their activities or they would not do so at all.

For instance, the head of a kitchen of a big hospital developed a system offering patients a large choice in composing their meals. The system also allowed patients on a special diet to make a selection (see illustration 1).

Diet for diabetes
DINNER ON THURSDAY, A SELECTION FROM:

first course
b. cream of garden herbs
d. meat stock

meats
b. veal steak
c. scrambled eggs
d. chicken breast

sauce/gravy
a. gravy
c. mild curry sauce
d. gravy of chicken

vegetables
a. French beans
b. asparagus
c. green peas
d. green beans

potato dish
b. potatoes, boiled
c. mashed potatoes
d. rice

side-dish
a. beetroot salad

dessert
b. raspberry custard
c. vanilla custard
d. low-fat yoghurt

page 8

non-diet meal
DINNER ON THURSDAY, A SELECTION FROM:

first course
a. onion soup
b. cream of garden herbs
c. Waldorf salad
d. meat stock

meats
a. Kromesky
b. veal steak
c. scrambled eggs
d. chicken breast

sauce/gravy
a. gravy
b. sauce(touch of port)
c. mild curry sauce
d. gravy of chicken

vegetables
a. French beans
b. asparagus
c. green peas
d. green beans

potato dish
b. potatoes, boiled
c. mashed potatoes
d. rice

side-dish
a. beetroot salad
b. apple sauce

dessert
a. rice tart and tutti frutti
b. raspberry custard
c. vanilla custard
d. low-fat yoghurt

page 8

Illustration 1.

By providing patients with choices, together with a keen buying policy he managed to reduce the amount of food that had to be thrown away from 40% to 5%. However, he said, there was no mathematics in his or his employees' daily work!

Similary an experienced carpenter working in the building industry pointed out that, perhaps, he did do some calculating once in a while but that he did not really come across any mathematics. He had never related all the measuring and marking he did to the (Euclidian) geometry of his former schooldays.

It was very difficult for the people who were interviewed to imagine what kind of mathematics is taught today: real world math. Further explanation did not help much: it was still difficult for them to distance themselves from the picture they have of schoolmathematics. Therefore we asked them to tell us about their jobs, observed them while at work and asked for further information if their stories or the work observed gave us reason to do so. We no longer stressed that we were in search of mathematics. This approach yielded a sufficient number of examples of mathematical activities on the shop floor to set to work designing teaching materials that meet the demand.

We also inquired about the requirements employees needed to meet (requirements of employees). The main things were that employees would have to be able to understand instructions without difficulty and carry them out without relying too much on others. They should also have a clear picture of the work as a whole, feel responsible, and be accurate. Hardly any of the skills explicitly taught at school were involved.

4. THE MATHEMATICS OF THE SHOP FLOOR

The mathematical skills of the future shop floor of learners with lower abilities can easily be divided into four groups. Each will be illustrated with the help of an example.

Situations strongly appealing to three-dimensional thinking. In hospitals employees sterilizing surgical instruments are expected to arrange those instruments under and over each other on trays as shown in a picture (see illustration 2). This, because it is essential for surgical assistants in the operating theatre to select the right instruments without having to look at them.

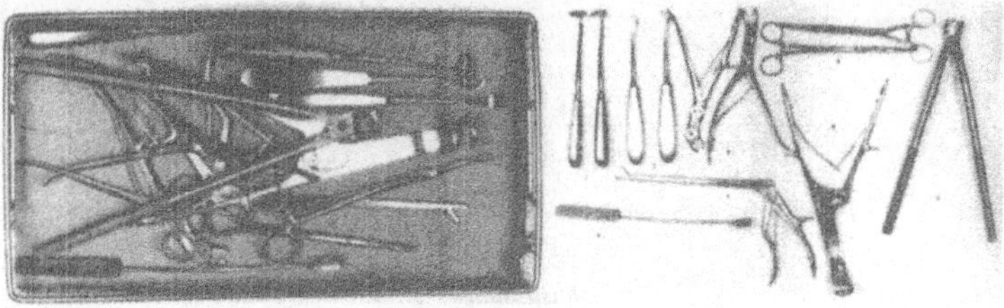

Illustration 2.

The use of (complicated) tables, manuals, diagrams and other information systems.

The maintenance engineer of the same hospital records the consumption of water with the help of 21 different meters in the water system. So far, using his records, various cuts in the consumption of water have been realized in practice, using his records. (See illustration 4)

Situations frequently requiring the use of amounts, sizes, prices and other units for calculating or measuring. To determine the right colour of pointing material on a building site, different proportions need to be tested. The composition of materials and colours selected needs to be maintained throughout the entire project to prevent differences in colour.

Different sets of rules and procedures determine the actions taken. These are of major importance in all three of the above-mentioned areas.

To calculate the distance between roofbattens requires the following procedure:

a.

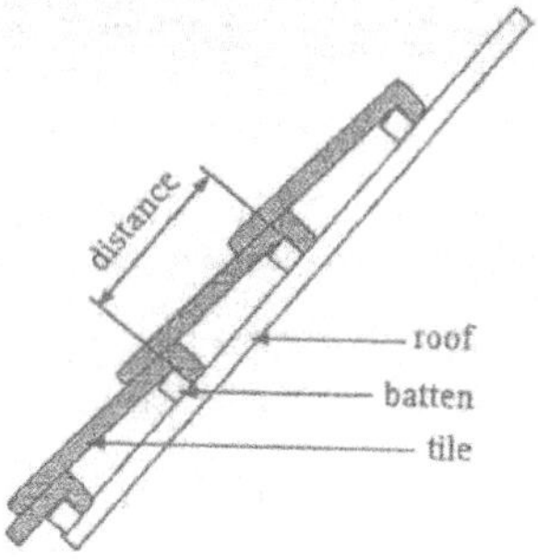

b. Take ten tiles.
c. Stretch the row of tiles as far as possible and measure the row.
d. Push the tiles together as far as possible and measure the row again
e. Take the average of the lengths measured.
f. Divide the answer by ten. The answer is the required distance between the battens.

Generally speaking, shop floor mathematics is elementary in character, non-formalized, and strongly related to necessity; a pragmatic use of mathematics. For example, a carpenter at work on a building-site will prefer using a spirit level when marking off lines. Should this not suffice, e.g. in the case of the horizontal laying off of a right angle, he will do this with the help of the 3-4-5 corner.

A number of topics in the current mathematics programme for learners with lower abilities finds few if any applications at all on the shop floor. Some examples are: linearity, stating a simple formula, drawing graphs, Pythagorean Theorem, properties of angles, square roots, statistics.

In contrast the situations on the shop floor requiring mathematical skills are fairly complex in character. Frequently, the skills are strongly integrated with each other and in the situation itself. For example, it is essential for a kitchen seller to be able to combine his three-dimensional visual imagination with the results of measuring and his ability to calculate. The whole of this should fit in with another complex system of information: some 40 pages of products the manufacturer of kitchens can deliver: base units, wall cupboards, grips, front types and colours, built-in equipment, etc., etc.

These examples help show why learners with lower abilities -and in fact all learners- experience difficulties in the application of even very simple mathematical skills: they are usually to be carried out within complex work situations.

5. THE TRANSLATION TO WORKSHEETS

As said above the writing and testing of teaching materials in classroom situations plays an essential part in answering the leading question of the project. To produce suitable teaching materials both the teaching strategy as well as the content had to be observed.

The teaching of learners with lower abilities often confronts teachers with problems related to students' low involvement to book learning and skills learnt at school. The origin of this attitude may be found in prior failure many of these pupils have experienced in schoolmath. "I can't do sums" or "I can't do math" are repeatedly heard among pupils in these groups. The academic style of book learning may be another stumbling block.

It is not without reason that the opening sentence of this article is: " What do I need maths for if...".

Furthermore, the number of pupils in this group with specific learning or behavioural problems is relatively high compared to other types of secondary pupils. Related to these aspects are problems in the fields of motivation and concentration that these pupils bring along, especially when they do not value the subject taught.

The translation of the above problems into teaching materials presented a number of conditions the teaching materials would have to fulfil (requirements of teachingmaterials). With respect to the mathematical content:

Useful, not formalized mathematics and even more, mathematics used on the shop floor.

Mathematics, that is perceptible and learnable by learners with lower abilities.

It turned out that these two conditions complemented each other well.

Quite a lot of shop floor mathematics is present in the curriculum for the first two grades in secondary education. By focusing on this, it would not be necessary for pupils to concentrate both on a new kind of mathematics and its application, but on the latter only.

With respect to the learning process:

A clear and acceptable role for the student (mostly a prospective professional worker)

Pupils carry out tasks serving a clear purpose.

Language usage suiting both the pupil's abilities and his\her task and role

Using the student's ready knowledge

Giving students time and facilities to explore the overall situation.

The role of the future employee was at the centre. Thus, the position and the use of the course content could be explained and it would also give the pupils status and responsibilities going far beyond that of just a pupil at school. In addition, the task set was related to a clear objective: a task derived from actual practice that has to be carried out according to certain standards.

Language usage had to be clear and unambiguous, encouraging independence, a co-operative attitude and the will to explore. To mention a single aspect: The pupil's task was not so much formulated in terms of a product but rather in terms of activities the students were expected to perform. This stimulated pupils to bring forward knowledge of their own and make use of the open character of the assignments.

Those who are familiar with the starting points of realistic maths education will recognise many of the conditions described above.

An extensive story was developed with respect to the two themes. The theme entitled 'A New Kitchen' tells the story of a young couple, Tim and Marita. It describes what they go through and have to think of once they have decided to buy a new kitchen that has to be installed in an existing house. Within the scope of this story a pupil sometimes plays the part of Tim or Marita but in most cases he\she is the expert assisting them in their process of decision-making concerning the design of the new kitchen and the execution of their plans.

Three of the many possible lesson series that can be introduced on the basis of this story line have been worked out within the project.

The same holds for the second theme, entitled: Behind the Scenes of the Hospital. In that case five lesson series have been worked out. This storyline is about a girl suffering from diabetes. While in hospital she meets different workers and professionals there, showing her some aspects of their work.

6. AN EXAMPLE.

A hospital is a vast consumer of water. An 800-bed-hospital needs as much as 2000 cubic metres a week. The water in this hospital does not come straight from the water supply system. Two large pure water cellars contain a 300 cubic metre buffer stock, which is enough for about one day. The water is needed in a great number of different places for a great variety of purposes.

In our example the consumption of water is recorded at 21 different spots.

The consumption of water was kept up to date with the help of a very complicated table (see illustration 4). This table did not just serve recording purposes. On the basis of the data the table provided, measures could be taken to prevent water being wasted. In practice the data in the table have also helped to locate failures in the system.

WATER CONSUMPTION COMPONENT FLOWS (m3).

meter	system	week 51.	week 52.	week .1..	week .2..
INTAKE					
04	SUPPLY PWC	1058	1059	1165	1166
05	SUPPLY PWC	867	867	971	971
	TOTAL INTAKE	1925	1926	2136	2137
REDUCTION SOFTENER					
06	softener up to	1161	1162	1393	1393
07	softened water	1137	1138	1352	1353
regeneration 6 - 7		24	24	41	40
tapwater cold: total in - 6		764	764	743	744
HOT-WATER SYSTEM					
01	HOT-WATER PUMP	87	87	78	79
02	HOT-WATER SYSTEM	0	0	0	0
	HWS. TOTAL	87	87	78	79
03	COOLING TOWERS	0	0	0	0
DEMI INSTALLATION					
08	COLUMN 1	48	49	13	14
09	COLUMN 2	40	41	14	15
10	CHEMICALS	1	1	1	2
11	DEMI TOTAL	11	11	10	10
REGENERATION 8+9+10-11		78	80	18	21

COMMENT
demi week 1 en 2 filter failure

METER	SYSTEM	week 51.	week 52.	week .1..	week .2..
INTAKE SOFTENER					
BASEMENT					
12	CH-INSTALLATION	0	0	0	0
13	STEAM GENERATORS	105	105	89	89
14	CHAMBER POT RINSING	466	466	795	796
15	BOILER KITCHEN	8	8	4	5
16	BOILER 13 3/4	79	80	75	75
17	BOILER 13 1/2	84	84	75	75
18	SWIMMING POOL FYSIO	-	-	-	-
6th FLOOR					
19	CH-INSTALLATION	0	0	0	0
2-P					
20	STEAM GENERATORS	-	-	-	-
21	CH-INSTALLATION	-	-	-	-

COMMENT
meter 14 low occupancy during festive season

Illustration 4. Table of component flows.

SOFT WATER

There are two water meters near the water softener. Meter 06 measures the amount of water the softener takes in. Meter 07 measures how much water actually passes the water softener.

45. Draw in the watermeters 06 and 07 in the graph showing you how the water flows

Not all water passing the water softener will be used as soft water. Part of the water disappears into the sewer. This aspect is referred to as regeneration.

46. How much regeneration is measured in week 51?

47. How did you calculate the answer to 46?
    Where in the table can you find this?

You can also calculate how much cold water comes in straight from the pure water cellar.

48. How much cold water was needed in week 51?

49. Write down the sum that helped you find the answer to 48.

50. Is there a watermeter to measure the amount of cold water used?
    Do you believe a water meter for cold water a necessary device?

51. Can you find any other sums in the table?
    If so, which ones?

Illustration 5. Worksheet

Prior to this example pupils will be asked to form a picture of the water system before looking at the table itself. At this stage they can also bring in everything they already know about the subject, e.g. by comparing their home situations to that of the hospital. Questions like: do we use soft water at home, too? Or, is there a carwash at the hospital? Answers are put down in a graph describing the differences between home and hospital. Then they use the information the table provides, detect the connection between data and select what is useful. Note, that here a clear change of roles of the pupils takes place. In the pupil's role the pupil first makes an assessment of the new situation. Using the information in the table, however, they are entirely absorbed in the role of the professional, responsible for the functioning of the water system. Once they start linking the data in the table to the graph representing the water system of the hospital, they return to their role as a pupil trying to connect newly gained knowledge with what they were familiar with already.

Trying out the teaching materials confirmed that the conditions that would have to be met in the development of teaching materials were well-selected. The authenticity of the situations presented turned out to be of great importance as quite a number of pupils actually did have some shop floor knowledge. This offered them the opportunity to bring in ready knowledge and skills.

The role of the teacher was a crucial one with respect to making time and creating opportunities for pupils to become familiar with the situation described. Information and instructions included in the teaching materials failed to incite pupils to do so, and, in consequence, did not elicit desirable behaviour. Furthermore, it became clear that extensive exploration of the situation was essential for pupils to be able to deal with the assignments. If teachers paid too little attention to this aspect, pupils ran a great risk of getting stuck in the teaching materials. And, characteristically for this group, many of them would then drop out. So, in addition to the teaching materials an extensive teacher manual was developed.

The experiments have also shown that students are very serious in carrying out their professional duties and that they feel very responsible to create a good product and achieve good learning outcomes. In all try-outs we never heard the question: 'What do I need maths for if I want to be...?'

7. INFLUENCE ON CURRICULUM DOCUMENTS AND EXAMINATION SYLLABI

The curricula and examination syllabi of all subjects taught in the lower and intermediate levels in secondary education have only very recently undergone large-scale changes. The following tendency can be observed with respect to the general subjects: Previously, the programmes were based on the academic content of the subjects and on the requirements for further education. Now the programmes are based on the student's ability and development as well as the demands made by society, further education and future professions.

This large-scale operation offered the opportunity to introduce the results of our project into the discussion on changes in the maths programmes.

With respect to the determination of the attainment targets, especially in case of the lower level, the actual use on the shop floor was of great importance. Some examples:

At the lower level Pythagoras' theorem is no longer included in the examination syllabus. However, explicit attention is paid to the application of equal proportions when drawing and estimating angles and distances (bear in mind the 3-4-5 corner).

Drawing up tables and making graphs are still part of the programme, though these activities hardly ever occur on the shop floor. Major arguments to do so were, firstly, the value of the independent design of objects for the learning process. Secondly, it is a fact that a skill like this one is a support to pupils whenever they will have to act in the position of consumer, volunteer or any other role in society.

This skill will help them compare information, test it, and be of support in forming an opinion and defining their position.

Formal notation and the use of variables and equations have virtually disappeared at the lower level. Formulas only occur in the form of word formulas and rules of thumb. The concept of equations is dealt with only in relation to graphs. Other concepts such as periodicity, interpolation, extrapolation and linearity as formal concepts are no longer included in the curriculum, which is also affected by the information gathered about mathematics on the shop floor.

At the level of general educational objectives, each of the general subjects is expected to contribute to the pupil's ability to reflect on the future. Part of this involves gaining insight into trades and occupations including actual practice and current developments within this and judging one's abilities and interests in the light of professions. It is another outcome of this project that the subject of mathematics can live up to this potential contribution to educational objectives.

8. FINAL CONCLUSIONS

The project has made it clear that the use of mathematics on the future shop floor of learners with lower abilities can be made visible in maths classes, al be it under well-defined conditions concerning content and teaching methods. To put it even stronger, it has become clear that contents responsive to the student's picture of the future, leaving out unnecessary formalizations and abstractions, have highly motivating effects on students and give them a feeling of responsibility. Thus, the outcome of the project presents suggestions for the development of maths teaching for learners with lower abilities in general, fitting their abilities and interests. To be perfectly clear, it has not been a subject of study to find out how maths education might contribute in a qualifying way, to the student's vocational preparation. The positive reaction from tradesmen, foremen and supervisors who were interviewed at the initial stages of the project to the materials developed (the teaching materials in particular) indicate that our approach has possibilities in this respect. The head of the sterilization department in the hospital is now using the teaching materials we developed for the vocational training of his staff.

9. REFERENCES

de Haan, J. (1989). What Do I Need Maths for if I Want to be a Maternity Welfare Worker? *(Dutch) Nieuwe Wiskrant* (May 1989), 40-44.

ten Hove, J., et al. (1996). A new kitchen (Dutch). *SLO, Enschede.*

de Lange, J. (1994). *Curriculum Change: an American-Dutch Perspective.* Paper presented at the 7th International Congres on Mathematical Education, Les Presses de L'Université Laval, Sainte-Foy.

Miow. (1994). *Basic Education in the Netherlands: the Attainment Targets.* Zoetermeer.

van der Zwaart, P. (1996). *Behind the Scenes of the Hospital (Dutch).* Enschede: SLO.

van der Zwaart, P. (1997a). Slow Learners, Mathematics and Future Profession. In S. K. Houston, et al. (Eds.), *Teaching and Learning Mathematical Modelling* (pp. 219-227). Chichester: Albion Publishing.

van der Zwaart, P. (1997b) *Mathematics and Work. Put Work into Mathematics.* (Dutch) Enschede: SLO.

10. NOTES

[1] This sector is referred as 'pre-vocational education' in the Netherlands.
[2] See: Zwaart, 1997 a, for the role of the lite carried out
[3] See: Zwaart, 1997 a.

CHAPTER 7

CLASSROOM TEACHERS DOING RESEARCH IN THE WORKPLACE

JOHN HOGAN AND WILL MORONY

Abstract. This chapter describes a methodology for using classroom teachers to gather information about mathematics in the workplace. It was used as one aspect of a project run by the Australian Association of Mathematics Teachers Inc (AAMT). The purpose of the project was to develop better understandings of what it means for a person to use mathematics when completing practical tasks. To help with this task snapshots of people at work were collected by having teachers, go into the workplace, shadow a worker for up to a day, interview the worker, then write up their findings. This chapter provides an overview of the project, an outline of the broader educational context, a description of the processes used and the methodology issues which emerged, some snippets from the research and how the teachers felt about the experience.

1. INTRODUCTION

The workplace has become a focus for much of the education and training curriculum for post-compulsory students in Australia — not only as a means of ensuring that students acquire the necessary knowledge, skills, and attitudes which prepare them for the workplace, but also because the workplace itself has, for an increasing number of students, become a site where the learning occurs. It now seems more important than ever to know how mathematics is located in the workplace and how workers use mathematics to help get their jobs done.

For many school teachers, mathematics teachers included, school has been their only workplace. Few have had extensive working experience out of school, let alone had the opportunity of studying the workplace to explore how mathematics is used for practical purposes and hence to consider the implications for the school curriculum and their own pedagogy.

A project conducted by the Australian Association of Mathematics Teachers (AAMT) in 1995-97 developed a way for school mathematics teachers to conduct research in the workplace and reflect on this experience for their own learning. The AAMT recommends the methodology as an affordable yet significant experience which should be made available to all teachers. A description follows.

Bessot & Ridgway (eds.), Education for Mathematics in the Workplace, 101—113.

2. BACKGROUND

The Australian Association of Mathematics Teachers is the pre-eminent Australian professional association in school mathematics education. It has a strong interest in national and state initiatives which support the improvement of teaching and learning in mathematics and the development of numeracy education in Australian schools. The Association's 6000 members come from all states and territories and all levels of schooling, and form an extensive network of committed and enthusiastic mathematics education professionals.

The AAMT has undertaken a number of research and development projects over the past few years, in addition to building its core business of providing professional reading, resources, programs, and events for teachers of mathematics. The national association is a federation of state and territory based teacher associations. These local Affiliated Associations are the agents of its national projects. Key 'Contact People' are identified in each of the affiliates and the project leader works, to a large extent, through these individuals. This adds to communication complexity, but enables truly national projects to flourish.

2.1. The Key Competencies

In the early 1990s, a national report (Mayer, 1992) of the Committee To Advise the Australian Education Council and Ministers of Vocational Education, Employment and Training developed a set of employment-related Key Competencies that were seen as

> 'essential for effective participation in the emerging patterns of work and work organisation...also...for effective participation in further education and in adult life more generally'*(Ibid, p, ix)*.

This committee concluded that there were seven Key Competencies: *Collecting, Analysing and Organising Information; Communicating Ideas and Information; Planning and Organising Activities; Working with Others and in Teams; Using Mathematical Ideas and Techniques; Solving Problems; and Using Technology.*

The key competency, *Using Mathematical Ideas and Techniques (UMIT),* as described on pages 33-35 of that report, was underpinned by the following major ideas:

-clarification of the purposes and objectives of the activity;
-selection of mathematical ideas and techniques;
-application of mathematical procedures and techniques;
-judgement of level of precision and accuracy needed; and
-interpretation and evaluation of solutions.

Between 1993 and 1996 the Australian Commonwealth Government funded projects in every state and territory in the education and training sectors to research how the Key Competencies might be integrated into curriculum, and to investigate the implications for assessment and reporting.

2.2. The RIUMIT Project

One such project, funded by the Department of Education, Employment, Training and Youth Affairs (DEETYA), was *Rich Interpretation of Using Mathematical Ideas and Techniques* (RIUMIT). The aim of this project was to promote an informed view of the UMIT competency to educators in all relevant areas and provide strategies for the effective development of this competency in young people.

The motivations for this project were many.

One was a sense that educators and employers using the Key Competencies often misunderstood, or avoided, the UMIT Key Competency. Misunderstood in that UMIT was equated with school mathematics and hence seen to be of a different category from the other key competencies, avoided because so many people view themselves as incompetent in the area of school mathematics.

Another was a widely held belief that people increasingly needed the ability to use mathematics in order to actively engage with a rapidly changing world. This situation was seen to be further complicated by advances in technology which were changing the nature of the mathematical ideas and techniques needed in the society of today and tomorrow.

A third was a lack of practical examples (of people using mathematics in work situations) in the literature to inform people's understanding of the Key Competency.

The RIUMIT project sought to involve a diverse range of educators (in terms of geography, education sector and area of interest) in both the research and development aspects. Research was conducted by posing questions, exploring and reaching conclusions about UMIT in the workplace, school and vocational, educational and training (VET) examples of UMIT, social justice issues surrounding UMIT, the related literature, key stakeholders' views of UMIT, and other Key Competency project reports. Materials that were developed to support wider dissemination and discussion of the findings included work stories, an information pamphlet, a video, and various professional development resources. The focus of this chapter is but one aspect of the above, that is, teachers doing research in the workplace.

3. THE WORKPLACE RESEARCH METHODOLOGY

The UMIT competency attempts to describe what a person does when using mathematics to assist in doing a practical task. A premise of the RIUMIT project was that if descriptions of the UMIT competency could be developed which exemplify how it manifests itself in practice, then educators, employers and workers would be better able to identify and develop it. Thus one focus of the research needed to be the work place, where the intent was to describe the actual work as it was being undertaken, and then to conduct an analysis from the perspective of how mathematics is used. A similar methodology was used by Baturo et al (1990) who investigated the mathematical demands of a number of unskilled occupations undertaken by people with intellectual disabilities. The research team developed rich descriptions of each worker's job by interview and observation within the work place. These descriptions were then analysed for evidence of the mathematics associated with the work place, the informal and formal mathematics identified in the work place and the metacognition associated with the mathematics. Their descriptions and ensuing analysis supported the power of such a methodology.

3.1. The aim

The aim was to generate about 40 stories. These were to be examples from settings such as small business, large business, industry, home, community, self-employment, volunteers and so on. Types of work such as hospitality, engineering, retail, agriculture, office and clerical, tourism, building and construction were to be included. Workers were to come from a range of backgrounds. The term 'work' was used in its broadest sense. It included both paid and unpaid work in, amongst other contexts, both domestic and community settings.

3.2. The plan

The main purposes for the workplace research were to develop a set of rich descriptions of a range of people operating in the work place in order to increase our understanding of how mathematical ideas and techniques are used in practical situations; to support our pursuit of the project outcomes; and to develop a methodology that would involve as many people from affiliated associations as possible. It was hoped that the methodology itself could be used by these associations when undertaking similar research.

The intention was to develop a description of what was happening in work places without imposing a predetermined view of mathematics and without having this view of mathematics corrupt the analysis of the stories. We also wanted to minimise peoples' 'fears' of mathematics to further reduce the interference with the process.

3.3. Data gathering methods

As described earlier, each of the AAMT's Affiliated Associations had a contact officer for the RIUMIT project. They were invited to find five members who were willing to participate as work place teacher-researchers.

Workers from a sample of work places around the country were shadowed for a half day by a teacher-researcher who was looking to describe the work situation. The worker was interviewed to explore issues that had arisen during the observation and to further develop a detailed description of the work. The aim of the interview was to identify what were routine problems, and what new and unfamiliar situations had occurred. The workers were encouraged to talk through the processes they use to make decisions about what to do in the routine and non-routine situations identified. The teacher-researcher could negotiate with the worker and employer to return for clarification purposes.

Following the visit, the teacher-researcher prepared a descriptive story with vignettes of how the worker dealt with one or two 'interesting' situations. The description and vignettes were accompanied by an initial analysis by the researcher. It had been intended to include one by the worker, but they generally endorsed the observations of the researcher (see later). The meaning of a set of events is subject to interpretation in a variety of ways so the intention was to portray multiple perspectives. The worker was given the opportunity to read and comment on what was written by the teacher-researcher.

We were keen to ensure that:

-workers and their employers agreed to the research process;

-both the employer and worker agreed to the public use of the example in project writings;

-workers had the opportunity to critique the description; and

-workers agreed to the description being used in future publications.

The school was to be paid the equivalent of the teacher replacement required to cover the absence of the teacher for the training session, the shadowing day, and the debriefing session.

In addition the teacher-researcher was paid $200AUS for the time spent on the project in addition to the above. If the work observation occurred in the teachers' own time then they were paid at a rate equivalent to that of that state's teacher replacement schedule.

3.4. The trial

A research methodology was trialed and refined in Western Australia in December, 1995 and January, 1996. The purpose of the trial was to clarify processes and prepare the information package for the teacher-researchers.

Four secondary mathematics teachers from an outer metropolitan senior high school (students aged 13 - 17) in Western Australia were involved in a trial of the process. After a short briefing they each spent a day in the work place, wrote up their findings, and then spent a half day debriefing their experiences. Two of these teachers, along with two new to the work, trialed the revised process again. As well as helping to refine and improve the research methodology the trial also ensured that we take seriously the following interrelated issues.

The placement of teacher-researchers

In the first stage of the trial it became apparent that locating the teacher-researchers and workers needed to be handled carefully. In this stage there were no restrictions as to where the teacher-researchers were situated and it was decided that the easiest way to set the exercise up would be for the School Industry Link Teacher to find each of them placements in local industry. Despite the fact that it had been stated that the workers could be doing any sort of work, this exercise reinforced how much people's understanding and feelings regarding mathematics intruded on the type of job to be shadowed. All placements ended up in large industries and all the jobs were middle management, professional positions. One personnel officer from a large mining and processing company commented that she had looked all over the site and found that the accountant was the only person who she could think of who did mathematics in his job.

In the second stage of the trial the teacher-researchers were asked to use their own network of relationships to locate their own worker to shadow. This enabled a much easier entree into the work place. It removed the need for 'middle persons', such as School Industry Link Teachers and industry personnel officers to be involved in negotiating the placement. It also assisted with the relationship between the worker and the researcher (see below). It was decided to use this method for the project across the country. The teacher-researchers found their own placements which were checked, in consultation with the project director, to ensure that a range of types of work and areas of work were achieved.

The relationship between the teacher-researcher and the worker

Not surprisingly, some workers were nervous about being watched and often devalued their own work. One worker said 'I'll just be doing the same thing over and over so you might as well go home now'. Some seemed to be somewhat intimidated by the presence of a mathematics teacher. One researcher, in the first stage of the trial, said 'People seemed frightened of having someone look over the shoulder. They seemed intimidated'. Clearly the process needed to ensure that the teacher-researchers took the time to build a relationship and that they needed to value the work being done by the worker.

Having the teacher-researchers find their own work sites through someone they knew (see above) was the first step in helping the relationship become 'easier'. In order to get some balance in the relationships, the researchers were asked to be up front and say that they (the researcher) were there to learn from them (the worker). Both of these aspects were included and emphasised in the second stage of the trial with, it seemed, good effect. One of the teacher-researchers involved in both trials wrote 'This time around was more relaxed, and more enjoyable'.

It was also felt that if there were opportunities for the teacher-researchers to help with some of the work and/or get the workers to teach them about what they do, that this reversal of roles would assist with the development of equality in the relationships, as well as improve the quality of the description of the work (ie leave in).

Selecting the teacher-researchers

It was important to get a range of teachers from different contexts involved. It seemed quite easy to get teachers to volunteer for the trial. There was a danger, though, that they could all end up as male secondary mathematics teachers. Contact officers of the Affiliated Associations were instructed to ensure that there was a balance between male and female researchers, that the researchers came from a range of locations and for there to be some who were not secondary mathematics teachers.

Industrial issues

There was, at the time, considerable industrial turmoil in education systems in the states and territories. It was necessary in some cases to clarify that this research work was not a local state government initiative which teacher unions had banned.

The focus of the shadowing

In the trial one work place set up a 'visit' and invited the researcher to talk to a variety of people throughout the day. Given the project's focus on detailed descriptions of people's work it was decided that the researchers needed to stay focussed on the one person.

Collecting the data

It was hard for the teacher-researcher to 'get inside people's heads'. One suggestion was to, if possible, become a participant observer. That is, to assist workers doing their work and/or to ask them to explain to the researcher how they do particular tasks. This had also been identified as contributing to the relationship between the researcher and worker. It was also reaffirmed that it was more important to document some things in depth than to get a sketchy overview of everything. Another suggestion was that the teacher-researcher should, at the completion of the observation session, take some time to clarify those elements about which more information was needed from the interview.

Safety issues

In some settings access to what people were doing was limited by safety constraints. This raised the issue of insurance for the teacher while in the workplace. Teaching staff were covered by insurance provided the work was seen to be part of discharging their duties on behalf of the school. Each of the contracts, if being completed by a teacher, therefore needed to be endorsed by the School Principal.

Training the researchers

It was clear from the trial that ideally all teacher-researchers should have a briefing and a debriefing. Due to the constraints of time, money, and geography this was not possible. In place of a training program a detailed information package was supplied to all researchers. This included a description of the task, clear instructions on keeping the shadowing focussed, notes on collecting the information, building relationships, two written examples from the trial, and the contracts that they would be required to complete. The contract for teacher-researchers included a clause which required them to debrief with a project officer and allowed for the possibility of returning to the workplace to gather more information.

4. OUTCOMES OF THE RESEARCH

4.1. The workplace stories

Thirty stories were completed. Twelve of these, along with a shortened version of the information package for teacher-researchers) are available through the AAMT website (www.aamt.edu.au). The example below provides snippets from one story to illustrate the format and content. The full story runs to some 9 pages. This meant that the first intention of the research - to develop rich descriptions of people at work - had been achieved

THE BAKER
by Andrew Butler

BACKGROUND

The Firm

The bakery specialises in the production of a wide variety of 'boutique' breads including olive loaves, sun-dried tomato loaves, seaweed bread sticks, marsala buns, baguettes and kibble rye bread. ...

Two full-time staff are employed, with up to five part-time or casual workers assisting production depending on volume of orders. Production generally commences around midnight with deliveries being made on a wholesale basis to several corner stores and delicatessens.

The Worker

K is the founding owner manager of the bakery, having previously worked as a self-trained chef in several leading restaurants. Prior to this *K* trained as a music teacher...

The Job

The regular processes performed by *K* involve taking daily orders and subsequent preparation of ingredients for the production of the various flavoured breads. The hand-mixing of dough is performed by the two full-time staff and individual loaves are prepared and baked by the other casual staff in addition to *K*. Following baking, *K* supervises the make-up of orders and delivery to retail outlets. *K* is also involved with ordering and receiving stock, invoicing, banking, reconciling accounts, training new staff, marketing and product development.

WHAT THE WORKER DID

K commenced her duties in the workplace at about midnight on Good Friday. The researcher arrived at 2.00 am! Easter Saturday. Prior to the researcher's arrival, the dough had been mixed and the yeast activated. Staff are beginning the process of mixing ingredients into the basic balls of dough. The quantities required have been determined by the daily orders taken by *K*.

The orders are tabulated on a spreadsheet like the one below, with the quantity of each type of bread ordered listed for each retail outlet. Rows and columns are totalled in 'dozens' of bread loaves.

VIGNETTE 4

***K** has prepared the basic dough mix using two medium bowls. The dough has risen and overflowed ('walked'). After using half of one bowl to prepare her 'smalls' and 'mediums', she is left with one and a half bowls to make two 'slabs'. She uses the following process to create two equal slabs out of one and a half bowls of dough. She cuts the full bowl into halves resulting in a total of three 'halves' when the other half full bowl is counted. She divides each half into thirds. This provides a total of nine 'thirds'. She uses four 'thirds' to create each slab. As she has a total of nine 'thirds' she has one remaining. She explains that this is enough quantity to be baked into an 'extra' loaf. (Is this the origin of the famous 'baker's dozen'?)...*

THE INTERVIEW

...

Question: What sorts of instruments do you use in measuring the quantities that you require?

***K**:* (Laughter) We work in dozens of loaves because that is how they are generally ordered by our customers. Because of that, each bowl of dough should make up one dozen loaves.

...

ANALYSIS 1: RESEARCHER

...

The instruction of the new employee, G, in Vignette 1 provided an interesting communicative insight between trainer and student. Rather than saying to divide the dough mixture into twelve equal proportions, ***K*** asked G to imagine the dough as a clock face and cut sections accordingly. As a result, G was quickly able to accurately estimate and fractionally divide the required proportions using the spatial 'mental map'...

WHAT THE WORKER SAID

***K**:* It all sounds very vague on reading it through, but I cannot stress enough that cooking in general is done more by 'feel' than by measurement, especially when dealing with a repetitive task. If it works, why change it?

As the production of bread began in a small way, just a few loaves a day, the measurements have developed because of what was available at the time (measuring implements). Now, with production having increased remarkably, to change the measuring system would be not only unwieldy but far too time-consuming. Teaching new people the measures hasn't been a problem.

4.2. Reflections by the teachers

The teachers involved talked of their experience in terms of their own professional development. They made comments about mathematics in the workplace, about doing the research, about the workplace itself and on how what they saw impacted their thinking about the classroom. The following observations have been made about the teachers' own reflections. (They do not include an analysis of the work stories or comment on how they inform how mathematics in the work place is enacted as this is not the purpose of this chapter.)

About their own feelings doing the research

The researchers certainly appreciated the opportunity to participate and described the experience as "very positive', "extremely interesting", "thoroughly enjoyable" and "professionally rewarding".

About mathematics in the workplace

Researchers were interested in how much mathematics they observed occurring in the workplace.

> [...] and it excited me to see how much mathematics a 'real world' worker used in his job.

> [...] the range of mathematical skills used by the worker - most of which he did not relate back to school mathematics and which he did not readily identify as mathematical skills.

They observed the way mathematics was used in the workplace. For example, observations were made about the level of accuracy in calculations which occurred in the different contexts. One work place had very liberal tolerances in their calculations and truncated everything, while another calculated everything to many decimal points.

About the workplace

Teacher-researchers talked of the insights into another workplace and the way things were done. Some of them were struck by the idiosyncrasies of work places where workers used their own way of doing things 'because they work'. It seemed that these workers didn't really understand why their methods worked. They observed that some workers had built up their own systems for doing things over a long time by trial and error.

> It was also an interesting experience for me to see, first hand, another side of the work force than education. This included some insights about differences in 'values and attitudes' as well as the work itself.

> [...] and appreciated a little more the complexities of peoples' work and their work environments.

> It was a pleasure to go to another person's workplace to watch them work; to see what they did, how they interacted with colleagues, to see if others work under the same tyranny of time-lines as teachers. It seemed a dignified and much gentler place. People were busy but not desperate.

About doing the research

It seemed that some began to appreciate the complexities and tensions of doing this kind of work.

> Qualitative research data gathering, particularly for this type of data, is time-consuming but rewarding in my experience. The unexpected can be observed and recorded for use, [...]
>
> During the observation phase I became more interested in watching and understanding S's work. I wanted to ask questions a lot but sensed that this would be a little disruptive.
>
> A lot of information and insight was also gained by questioning during the shadowing period and also the post-observation interview.
>
> Before I began this work shadowing assignment, I felt a little uneasy about the process, as I was entering someone else's domain. [...]
>
> My presence in the workplace stimulated conversation about people's perceptions of mathematics used in that workplace. This in itself was worthwhile, in that it may have increased some awareness of using mathematics [...]

About their classroom practice

Some of the teacher-researchers have begun to ponder how their observations in a work place might impact on their own classroom practice and the learning experiences that they provide for their students. For example, some were interested in the way problems were tackled and solved.

> [...] made me realise the variety of situations that [...] may be faced with. The problem-solving techniques that V has acquired have been invaluable in many of the situations that she is faced with. Perhaps more of these strategies should be emphasised in secondary education.

A number thought it would help answer students' questions in class like 'Why are we doing this?'.

> Satisfying as well - as I have often taught this to students who have said 'What is the use of this?'
>
> [...] I answered the 'Why are we doing this?' question with the standard replies, so it was very refreshing to hear R echo those remarks, but with some 'technical back-up'.

Some began to be aware of the dangers of not only taking the mathematics out of context and making use of the mathematics without regard for its context but also of misrepresenting the context.

> Although I can say that much of what A and G do is mathematics, it is imperative before these contexts are used in teaching situations that they be discussed with the communities concerned. These contexts have developed from other areas of knowledge and to put them into the basket of mathematical knowledge needs discussion. The reason for learning the skills could be lost if it was formalised and put into the classroom.

Some began to reflect on how people do their work in the workplace compared to how students do their work in school.

> It would be better to teach fewer things properly than to try and cover too wide a range of mathematical ideas.
>
> It seemed to me that the workers were engaged for about one third of their time in getting on with others, one third of their time in solving new problems and one third in interpreting information. It seemed to me that I could break up my lessons in class in the same way.

Another researcher noticed the importance of 'the ability to follow a set, standard procedure' but also that 'the ability to modify the procedure to suit individual preferences, while ensuring that the job is done properly, is also important.'

Clearly such comments depend as much on the workplace visited by the researcher as it does on the world view of the researcher. It is also clear from many of the comments that the teachers involved were engaged in significant learning about the workplace and the role of mathematics within it. this learning has demonstrably resulted in significant professional development in terms of approaches to the teaching and learning of mathematics.

5. CONCLUSION

This aspect of our work achieved much more than expected, particularly in the area of learning for the teachers involved. Not only did the project gather interesting descriptions of people at work to exemplify the key competency, but the methodology itself proved to be excellent professional development for the teachers doing the research.

One coordinator of a college/industry link program has planned to adopt the idea

> [...] From this perspective (the idea of shadowing) I will be able to gain a greater understanding about the mathematics our Aboriginal students require to be able to operate in a particular work placement. The idea of shadowing will enable me to explicitly target the mathematical ideas and techniques needed for students to successfully engage in work placements and inform the respective teaching programs and workshops before they are placed in new situations.

A tertiary mathematics educator has talked of using the idea for assignment work for her student teachers. The methodology seems to have appeal for wider application other than just for mathematics teachers.

The stories themselves have proven of great interest to teachers, particularly mathematics teachers and adult numeracy teachers. They have been used in a range of professional development activities around Australia and have stimulated good discussion about mathematics in the workplace and how it might impact on the curriculum and classroom practice.

There were a number of constraints operating on our processes. These included the ambitious scope of the project overall, the limited time available, the geographic location of the researchers and the number of researchers. All of these impacted on the methodology chosen, the quality of the work and resulted in some variation in the way the work was carried out, and the time it took from start to finish. However there seemed to be significant benefits from the scheme in that teacher participation was very affordable and manageable. The fact that teachers were still willing to participate in times of industrial turmoil was indicative of their interest in the possibilities of such work and the importance they saw in doing it.

6. APPENDIX: EDUCATION IN AUSTRALIA

The provision of public school and technical education is the responsibility of the eight state and territory governments in Australia. While mathematics has a core and obvious place in schooling, it is most commonly taught and learnt in a 'service' mode in technical and further education. Broad national priorities and directions are agreed at government level from time to time and these attract additional funding from the federal government. The work on Key Competencies covered in this chapter was one of these priorities. Also relevant is that there has been, in the time since this project concluded, an emphasis on Vocational Education and Training (VET) in schools. Typically, students have moved from school into either a technical or a higher education stream. Policy and programs have for some years sought to blur distinctions between the roles of schools, technical education institutions and universities.

Acknowledgments
This project was conducted by the Australian Association of Mathematics Teachers Inc, and supported by a grant under the Key Competencies Program from the Commonwealth Department of Employment, Education, Training and Youth Affairs.
The views expressed are those of the writers. They do not necessarily reflect the views of the AAMT or the affiliated associations and do not necessarily represent the views of the Commonwealth Department of Employment, Education, Training and Youth Affairs.
The writers are indebted to the work done by each of the state and territory AAMT Affiliated Associations.

7. REFERENCES

Baturo, A., Cooper, T., Walter, R., Waterford, P., and Watters, I. (1990). *Mathematics in the Workplace: Case Studies of Adults with Intellectual Disabilities*. Queensland University of Technology, Queensland: Project supported by Queensland Department of Employment, Vocational Education, Training and Industrial Relations, Centre for Mathematics and Science Education.

Mayer, E. (1992). *Key Competencies*. Canberra, AGPS: Report of the Committee to Advise the Australian Education Council and Ministers of Vocational Education, Employment and Training.

CHAPTER 8

MAKING AUTHENTIC MATHEMATICS WORK FOR ALL STUDENTS

SUSAN L. FORMAN AND LYNN ARTHUR STEEN

Abstract. In an attempt to improve education for all students, U. S. educators and employers have independently developed voluntary national standards for academic subjects and for various occupational skill clusters. These standards pose a special challenge to mathematics: to develop a core curriculum suitable for all students that is grounded in authentic, concrete tasks in which important mathematics is embedded in meaningful and realistic problems from life and work.

Historically, education in the United States, like Caesar's Gaul, has been divided into three tracks—academic, vocational, and general—that differ greatly in goals, status, size, and outcomes. Although two out of three U.S. secondary school students say that they intend to go to college, and sixty percent do begin full time postsecondary study, only one-third actually complete a college-preparatory program while in high school (National Center for Education Statistics (NCES), 1992). In contrast, only about fifteen percent of high school students pursue a vocational program, and only half of them (approximately eight percent) complete it (NCES, 1996). The remaining students receive a 'general' diploma that does not prepare them either for the world of work or for higher education.

Not surprisingly, the academic track is the locus of prestige for students, parents, and teachers. In recent years, political pressure has aimed the spotlight of academic accountability on calculus as a metaphor for educational quality. Although calculus has historically been principally a postsecondary subject in the United States, it is now a chief public indicator of high quality secondary schools. Intense public pressure to measure schools by the number of students who earn credits in the Advanced Placement (AP) calculus program has effectively lowered the status of most applied and vocational options. Even the National Collegiate Athletic Association (NCAA), which regulates scholarships for college athletes, is now unwilling to certify high school mathematics courses that are deemed too applied or vocational.

With all this emphasis on preparation for college, neither parents nor politicians pay much attention to preparation for work. In contrast to other nations, the United States has no tradition of high-quality apprenticeship education for students in the

Bessot & Ridgway (eds.), Education for Mathematics in the Workplace, 115—126.

final years of secondary education. Most states require secondary school students to complete certain courses (counted in so-called 'Carnegie units') that generally do not allow credits for experiential learning. Although some vocational programs do include work-based (or 'co-op') components, most rely on school-based exposure to vocational subjects such as automotive repair or carpentry. Being isolated from current industrial practice, these school-based programs quickly become outdated. Without a continuing commitment to cooperation, it is very difficult for teachers, curricula, and textbooks to keep pace with the rapid changes and increasing rigor of the contemporary high-performance workplace.

These two handicaps—operating in the academic shadow of calculus and lacking the rigor of contemporary industrial requirements—burden vocational education in the United States with low status that seems very difficult to eradicate, notwithstanding evidence of high-paying technical jobs in many industries. Indeed, increasing demand created by worldwide economic trends (e.g., ISO standards, international subcontracting, and information technology) combined with an educational system that is out of step with the needs of the modern workplace have created a severe shortage of qualified technical workers in the United States (Marshall & Tucker, 1992; Judy, 1997). None of the three educational tracks commonly offered by U.S. secondary schools prepares students to meet the challenges of careers in the modern international economy.

Postsecondary Enrollments (in Thousands):	*1990*	*1995*
Arithmetic & Basic Skills	224	252
Elementary Algebra	330	360
Intermediate Algebra	431	409
College Algebra & Trig	597	663
Elementary Statistics	171	236
Other Precalculus Courses	384	405
Secondary-Level Subtotal	2137	2325
Calculus & Linear Algebra	776	668
Computer Science	278	142
Advanced Statistics	52	44
Other Advanced Courses	120	96
College-Level Subtotal	1226	950
Total Postsecondary	3363	3275

Table 1. Postsecondary U. S. mathematics enrollments.

Surprisingly, much the same can be said about preparation for postsecondary education. In fact, weak mathematics is the most common source of inadequate preparation for college and university programs. It is not uncommon in the United States for secondary school graduates to begin their postsecondary mathematics with elementary or intermediate algebra, or even arithmetic (see Table 1; Loftsgaarden, Rung, & Watkins, 1997). Moreover, a recent survey (National Association of

Manufacturers, 1997) reveals that more than half of current manufacturing employees lack basic skills in mathematics and in document literacy (e.g., reading diagrams, graphs, and flowcharts).

1. THE STANDARDS MOVEMENT

Educators and employers responded to these problems by creating standards for academic subjects and for occupational skills. The movement in the United States for academic standards began in mathematics with the publication of *Curriculum and Evaluation Standards for School Mathematics* (National Council of Teachers of Mathematics (NCTM), 1989). Subsequently, K-12 standards have been developed in history, language arts, science, fine arts, foreign languages, and other subjects, as have standards for the first two years of college mathematics before calculus (American Mathematical Association of Two-Year Colleges, 1995).

Employment standards were first expressed in a widely-read government report, *What Work Requires of Schools* (Secretary's Commission on Achieving Necessary Skills (SCANS), 1991). This influential report outlines five competencies (using resources, information, systems, interpersonal skills, and technology) built on a three-part foundation (basic skills, thinking skills, and personal qualities) required of any employee in any job. Subsequently, more focused occupational skill standards were developed for entry level positions in two dozen different employment clusters ranging from industrial laundering to photonics and from retail trades to advanced high performance manufacturing. (Appendix A contains a complete list of these clusters; related Web sites are listed in Appendix B.)

Not surprisingly, the word 'standards' means different things to different people. For NCTM, a standard expresses a vision about what is valued—a banner behind which to rally support for improvement in mathematics education. For industry, standards specify the knowledge and competencies required to perform successfully in a specific occupation (Bailey, 1997). For the public, standards provide both a goal (what should be done) and a measure of progress toward that goal (how well it is done) (Ravitch, 1995).

Both sets of standards—academic and occupational—are voluntary. Autonomy of schools and businesses from governmental authority is an abiding value of the American political system. Just the idea of national 'standards' is enough to make many individuals suspicious because externally imposed mandates are seen as an unconstitutional intrusion on the independence of local communities. To coordinate the emerging occupational skill standards, in 1995 the United States Congress established the National Skills Standards Board (NSSB). However, in the face of widespread criticism about federal intrusion in education, the Congress rescinded a parallel effort to coordinate or promote academic standards.

In addition to being voluntary, the two sets of standards, like the academic and vocational tracks in the schools, are totally independent. Educators developed academic standards with virtually no input from employers; business and industry developed occupational skill standards with virtually no input from educators. Thus, for example, what the two sets say about mathematics is very different. NCTM recommends a mathematically rich eleven-year curriculum focused on problem solving, communication, reasoning, and making connections; the occupational skill standards, by and large, call only for lower-level mathematical skills that are normally covered at the transition from primary to secondary education. These different visions of essential mathematical skills, amplified by economic analysis (Murnane & Levy, 1996), send mixed messages to teachers and schools.

Although the academic and occupational standards address different issues and outline different expectations, they are surprisingly consistent in their vision of effective pedagogy. All the standards argue or imply that learning is enhanced when embedded in rich, authentic contexts; when students engage with each other and with the world around them; and when students are expected to experience, explore, and explain. Both the occupational and academic standards clearly recognize the limitations of traditional textbook learning—what Lee Shulman (1997) has termed the challenges of amnesia (loss of learning), illusory understanding (idols of learning), and inert ideas (uselessness of learning). The standards endorse, either implicitly or explicitly, Shulman's prescription for effective learning: activity, reflection, and engagement.

Finally, despite the enthusiasm of their developers and supporters, neither set of standards has met with widespread approval. School mathematics standards have been intensely controversial, with politicians and educators taking sides through newspaper editorials and political wrangling over such issues as basic skills, mathematical rigor, calculators, and contextual problems. (Similar arguments have plagued the standards in other academic subjects, especially history and language arts.) In contrast, the occupational skill standards have been largely ignored, or greeted with apathy. Few educators know anything about them, as do few managers in industry. As a consequence, occupational skill standards have stirred neither public interest nor passion.

2. STATUS, EQUITY, AND TRACKING

The goal of both educators and employers is to increase the quality and rigor of students' school experiences to ensure that graduates are well prepared for work and for postsecondary education. Although many schools view this as a choice—either work or college, either vocational programs or academic programs—many educators and politicians argue that all students should leave secondary school well prepared for *both* work and further education:

-Most of today's careers require some form of postsecondary preparation.

-Virtually all employees must expect to periodically update their knowledge and skills through some form of postsecondary study.

-Many students graduate from college with little knowledge of or commitment to any career or vocation; most would benefit from an early introduction to the world of work.

-Class and racial inequities in U.S. society are magnified by educational tracking that holds some students, primarily poor or minority, to lower standards than others.

NCTM argues in its *Standards* that all students should master a common core of mathematics. Separate tracks for students with different interests could achieve this goal, but only if each maintains a similar set of performance expectations. What is not acceptable, in this view, is a vocational or general track that sells students short by expecting little and accepting even less. Such programs, still too common in U.S. schools, offer the least education to students who have the greatest need.

To meet these challenges mathematics educators need to find a way to coordinate (or integrate) the mathematics and occupational skill standards so that students leave secondary school mathematically prepared both for postsecondary education and for the high performance workplace. However, unless parents, teachers, and students perceive that the mathematical expectations of what is now often called 'applied academics' is of comparable challenge and rigor to that of traditional academic programs, preparation for work in the schools will never achieve the same status as preparation for college or university.

3. BEYOND EIGHTH GRADE

One reason mathematics for work is held in such low esteem among parents and teachers is the perception that employers expect primarily basic computational skills. To be sure, they do expect basic skills first; this is the unmistakable message both of the SCANS report, the occupational skill standards, and public opinion (Wadsworth, 1997). But beyond basic skills, business and industry expect three kinds of mathematical performance rarely taught in the schools:

-Problem solving that relies on sophisticated uses of relatively elementary mathematical tools (Forman & Steen, 1995). In contrast to typical secondary school mathematics problems that consist primarily of routine manipulations involving advanced topics such as trigonometric identities and quadratic equations, these kinds of problems are more concrete and more common, yet just as challenging.

-Modern management tools such as systems analysis, PERT charts, and statistical quality control (Mathematical Sciences Education Board (MSEB), 1995. In contrast to typical secondary school topics such as factoring, finding roots, and solving equations that are included as preparation for the study of calculus, these other tools would prepare students to work in tomorrow's world.

-Mathematical habits of mind that are embedded in contexts such as measurement, management, and quality control (Packer, 1997). In contrast to the easily recognizable mathematical skills of arithmetic, geometry, and algebra, these distinctively mathematical patterns of thought underlie many aspects of the occupational skills standards but are rarely identified as *mathematical skills*.

Although many parents and teachers in the United States do not yet recognize a need to include these kinds of mathematical tasks in the traditional curriculum, others agitate for transformation (or at least enrichment) of secondary school mathematics to include more authentic tasks that reflect the ways in which mathematics arises in life and work. For some the goal of such a transformation is primarily vocational (to ensure that students are better prepared for work), while for others the goal is mostly pedagogical (to motivate more students to learn high quality mathematics).

Two current initiatives seek to advance this agenda. The U. S. National Academy of Sciences recently published a collection of essays and mathematical tasks suggesting ways to connect secondary school mathematics to the kinds of challenges students will likely encounter at work and in their daily lives (MSEB, 1998). And the National Center for Research in Vocational Education (NCRVE) sponsored a workshop entitled 'Beyond Eighth Grade' to highlight the rich mathematics embedded in workplace tasks and to strengthen the case for both rigorous academic standards in vocational education and authentic workplace applications in academic programs (Forman & Steen, 1997).

These initiatives identified several fundamental questions concerning secondary school mathematics, including:

- -Should students be encouraged to spend time on newer topics more in demand by industry (e.g., statistical quality control, spreadsheets) even if it means delaying the study of calculus?
- -How can mathematics teachers who have never worked outside the schools be expected to teach authentic workplace applications with competence and confidence?
- -Can a single mathematics curriculum provide both what industry wants of high school graduates and what higher education expects of entering students?
- -Are there many authentic workplace applications of (upper) secondary school mathematics?

The last question, which is really a prerequisite to all the others, is surprisingly difficult to answer.

4. AUTHENTIC TASKS

Mathematics is known by the problems it keeps. Children's homework reminds parents of their own mathematics homework. Each generation's successes (and traumas) are passed on to the next. For many, algebra is a rite of passage. Thus old problems recur like clockwork in textbooks and on tests. In the United States, entrance examinations for postsecondary education are filled with predictable problems:

- -A bag contains 28 pounds of sugar which is to separated into packages containing 14 ounces each. How many such packages can be made?
- -The sum of the first three of six consecutive integers is 27. What is the sum of the last three?
- -If the length of a rectangle is increased by 20 percent and the width is decreased by 20 percent, by how much is the area changed?
- -What percent of 1/2 is 3/4?

So too are exams that are required of prospective secretaries, postal workers, or plumbers:

-Find the next two numbers in the series 27, 26, 24, 23, 21, 20, 18, ...

-If C = (5/9)(F - 32), what is C when F is 50?
-30% of 420 employees in a department are clerks and 1/7 are typists. What is the difference between the number of clerks and the number of typists?
-If 12 word processing center employees can produce 12,000 documents in 20 days, how many documents can 18 word processors produce in 30 days?
-Mark is now four times as old as his brother Stephen. In one year Mark will be three times as old as Stephen will be then. How old was Mark two years ago?

Virtually no one ever encounters problems like these except in mathematics courses, yet they remain the staple of entrance examinations both for employment and for higher education.

At work, employees are much more likely to encounter problems such as the following, which are drawn from discussions with real people working at real jobs:

Ordering Drapes. A customer asks a sales associate in a large department store for advice about ordering draperies for a living room. The customer has a photograph from a magazine of the desired window treatment and a sketch of the room including locations of the windows and height of the ceiling. The salesperson needs to advise the customer on a number of issues. Will the treatment translate well to the customer's home? Is the scale of the room (and of the windows) appropriate? How can they take into account a difference in ceiling height? If the desired treatment makes sense in the customer's room, then the sales person needs to determine how much material needs to be ordered.

Arranging a Stockroom. A typical shoe store has a crowded stockroom in which several thousand boxes of shoes are stored. The boxes are labeled by manufacturer, style, color, and size, and are stacked floor to ceiling on rows of shelves. How should the shoes be arranged in the stockroom? Obvious options are by manufacturer, by style, by size, by frequency of demand, or by date of arrival. The main concern is getting shoes out to customers quickly and minimizing the amount of time spent returning shoes to their correct location once a transaction has been completed. But other concerns need to be considered as well. For example, when replacement stock or new styles arrive, clerks need to add the shoes to the stockroom in the appropriate locations. And clerks need to check the inventory weekly to verify records of change due to sales and receipt of new shoes as well as to identify misshelved boxes.

Machining Parts. In most manufacturing industries, machinists operate precision cutting and drilling tools to fabricate parts to designers' specifications. Often these machines are controlled by computers, but the machinist must provide the computer with the proper calculations. For example, a block to be machined sits on a compound sine plate five inches square, as shown in Figure 1. Determine the tilt of the upper and lower plates, and the number of inches of blocking required to achieve this tilt in order to level the block.

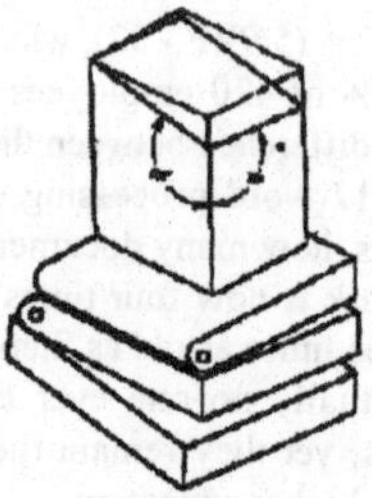

Figure 1.

Scheduling Elevators. In some tall buildings all elevators can travel to all floors, while in others certain elevators are restricted to stopping only at certain floors. Under what circumstances is it advantageous to have elevators that travel only to certain floors? Does it make a difference if the building is a hotel, an apartment building, or an office building?

Maintaining a Pool. The chlorine used to control microorganisms in a swimming pool dissipates in reaction to bacteria and the sun at a rate of about 15% per day. The optimal concentration of chlorine is between 1 and 2 parts per million (ppm), although 3 ppm is safe for swimming. How much chlorine should be added each day to maintain an appropriate concentration? Could one get by with a regimen of adding chlorine only once per week? What about twice a week?

Leasing Space. A small start-up company needs to lease space for its staff and operations. Available sites differ in such factors as gross square footage (on which the rent is based), usable square footage (omitting hallways, stairways, bathrooms, elevator shafts, and other service space); outside window space; and compactness (the ratio of perimeter to area, which indicates how spread out the office space is). This information needs to be calculated from floor plan drawings and then organized in a manner that will permit easy comparison.

These kinds of authentic mathematics-rich tasks—embedded in real contexts, mixing hard and soft data, amenable to multiple approaches and conclusions—are typical of life and work. Although they don't look like typical mathematics problems, they serve as effective Trojan horses for all the basic skills of school mathematics: (e.g., numbers, equations, graphs, measurement, geometry). At the same time, these tasks give students much needed experience in the more subtle cognitive skills required to employ mathematical thinking in the analysis of contextually rich tasks.

5. CHALLENGES

School mathematics in the United States builds on a long tradition of tracking which ill-serves many students. This tradition is being challenged by new academic standards that set high expectations for all students and by occupational skill standards that express what students need for entry level employment. While these standards have some important elements in common, they do not yet convey a unified vision—especially not of secondary school mathematics.

By seeking elements important to both sets of standards one can, however, discern criteria for secondary school mathematics that could form the basis for a unified core curriculum that serves both aims (Forman & Steen, 1999):

-Beyond arithmetic, all students should take a three-year core of mathematics consisting of statistics, geometry, and algebra in roughly equal proportion. These subjects represent the three great strands of mathematics—numbers, space, and symbols—as well as the basic mathematical needs of citizens for life and work.

-In this three-year core of secondary school mathematics, students should work extensively on authentic, concrete, contextual problems that require sophisticated multi-step reasoning with elementary mathematical tools. Such tasks will motivate students to master the use of important basic skills before moving on to more advanced and abstract topics required for postsecondary education.

-Students who complete this core can then choose among advanced electives based on their interests. One option would be a precalculus course that introduces the algebraic manipulations and elementary functions that form the foundation of calculus. Another option would be advanced computer-assisted geometry (e.g., CAD/CAM) or statistical tools for quality control (SQC).

A core curriculum adhering to these principles, rooted in credible and concrete problems requiring only elementary tools, will make high school mathematics work for all students.

6. APPENDIX A: OCCUPATIONAL SKILL STANDARDS

Administrative Support	Professional Secretaries International, Kansas City, MO
Agricultural Biotech	National Future Farmers of America Foundation, Alexandria, VA
Automotive Technicians	National Automotive Technicians Education Found., Herndon, VA
Bioscience	Educational Development Center, Newton, MA
Chemical Technicians	American Chemical Society, Washington, DC
CADD	Foundation for Industrial Modernization, Washington, DC
Construction	Laborers AGC Education & Training Fund, Pomfret Center, CT
Electronics	American Electronics Association, Washington, DC
Electronics Technicians	Electronics Industries Foundation, Washington DC
Electrical Construction	National Electrical Contractors Association, Bethesda, MD
Forest & Wood Products	Foundation for Industrial Modernization, Washington, DC
Groceries	National Grocers' Association, Reston, VA
Hazardous Materials	Center for Occupational Research and Development, Waco, TX
Health	Assoc. for the Advancement of Health Education, Reston, VA
Health Care	Far West Laboratory for Educ. R&D, San Francisco, CA
HVAC	Vocational-Technical Educ. Consort. of States, Decatur, GA
Hospitality &Tourism	Council on Hotel, Restaurant & Institutional Educ., Washington, DC
Human Services	Human Services Research Institute, Cambridge, MA
Industrial Laundering	Institute of Industrial Launderers, Washington DC
Manufacturing	National Coal for Advanced Manufacturing, Washington DC
Metalworking	National Tooling and Machining Assoc, Fort Washington, MD
Photonics	Center for Occupational Research and Development, Waco, TX
Physical Education	National Assoc for Sport and Physical Education, Reston, VA
Printing	The Graphic Arts Technical Foundation, Sewickley, PA
Retail	National Retail Federation, Washington DC
Welding	American Welding Society, Miami, FL

7. APPENDIX B. SELECTED WEB SITES

Mathematics and Science Education:	
Amer. Math. Assoc. of Two-Year Colleges (AMATYC)	www.amatyc.org
Educational Resources Information Center (ERIC)	www.accesseric.org
Eisenhower National Clearinghouse (ENC)	www.enc.org
EXTEND: Broadening Reform in Mathematics Education	www.stolaf.edu/other/extend
National Council of Teachers of Mathematics	www.nctm.org
U. S. Government Offices:	
National Center for Education Statistics (NCES)	www.ed.gov/nces
School-to-Work National Office	www.stw.ed.gov
U.S. Department of Education	www.ed.gov
U.S. Department of Labor	www.dol.gov
Occcupational Skill Standards:	
American Society for Agricultural Engineers (ASAE)	www.asae.org
Center For Occupational Research and Devel. (CORD)	www.cord.org
Education Development Center (EDC)	www.edc.org
Electronics Industries Association (EIA)	www.eia.org
Graphic Arts Technical Foundation (GATF)	gatf.lm.com
Nat'l Center for Research in Vocational Educ. (NCRVE)	vocserve.berkeley.edu
National Center on Education and the Economy (NCEE)	www.ncee.org
National Coalition for Adv. Manufacturing (NACFAM)	www.bmpcoe.org/nacfam
National Retail Federation (NRF)	www.nrf.com
National Skill Standards Board (NSSB)	www.nssb.org
Business and Industry:	
Business Coalition for Educational Reform (BCER)	www.bcer.org
Committee on Economic Development	www.ced.org
National Alliance of Business (NAB)	www.nab.com
Reports and Publications:	
Advanced High-Performance Manufacturing:	www.bmpcoe.org/nacfam/skilstd1.html
Chemical Technicians Skill Standards:	www.acs.org/pafgen/slreact/environ/2-3e.htm
Crossroads in Mathematics (AMATYC):	www.richland.cc.il.us/imacc/standards
Curr. Standards for School Math. (NCTM)	www.enc.org/online/NCTM/280dtoc1.html
National Health Care Skill Standards:	www.fwl.org/nhcssp/health.htm
National Printing Skill and Knowledge Standards:	www.gatf.lm.com/skills.html
National Skill Standards for Electronics Technicians:	www.eia.org/cema/ps/res/nss.htm
Nat'l Skill St'ds for Hospitality & Tourism:	www.access.digex.net/~alliance/skills.html
Photonics Education:	www.spie.org/photonics_ed.html
SCANS Competencies:	www.stolaf.edu/other/extend/Resources/scans.html

8. REFERENCES

Mathematics: Standards for Introductory College Mathematics before Calculus. Memphis, TN: Author.

Bailey, T. R. (1997). Integrating Academic and Industry Skill Standards. National Center for Research in Vocational Education. Berkeley, CA: University of California.

Commission on Skills and the American Work Force. (1990). America's Choice: High Skills or Low Wages. Rochester, NY: The National Center on Education and the Economy.

Forman, S. L., and Steen, L. A. (1997). Beyond Eighth Grade: Report of a Workshop on Mathematical and Occupational Skill Standards: available at www.stolaf.edu/other/extend/Resources/beg.html.

Forman, S. L., and Steen, L. A. (1995). Mathematics for Work and Life. In I. M. Carl (Ed.), Prospects for School Mathematics: Seventy Five Years of Progress (pp. 219-241). Reston, VA: National Council of Teachers of Mathematics.

Forman, S., L., and Steen, L., A. (1998). Beyond Eighth Grade: Functional Mathematics for Life and Work. Berkeley, CA: National Center for Research in Vocational Education (MDS 1241).

Judy, R. W. and D'Amico, C. (1997). Workforce 2020: Work and Workers in the 21st Century. Indianapolis, IN: Hudson Institute.

Loftsgaarden, D. O., Rung, D.C., and Watkins, A. E. (1997). Statistical Abstract of Undergraduate Programs in the Mathematical Sciences in the United States: Fall 1995 CBMS Survey. Washington, DC: The Mathematical Association of America.

Marshall, R., and Tucker, M. (1992). Thinking for a Living: Work, Skills and the Future. New York, NY: Basic Books.

Mathematical Sciences Education Board. (1995). Mathematical Preparation of the Technical Work Force. Washington, DC: National Research Council.

Mathematical Sciences Education Board. (1998). High School Mathematics at Work. Washington, DC: National Academy Press.

Murnane, R., and Levy, F. (1996). Teaching the New Basic Skills: Principles for Educating Children to Thrive in a Changing Economy. New York, NY: The Free Press.

National Association of Manufacturers. (1997). The Skilled Workforce Shortage available at www.nam.org/Workforce/survey.

National Center for Education Statistics. (1992). National Education Longitudinal Study: Second Follow-up and High School Transcript Files. Washington, DC: US Department of Education.

National Center for Education Statistics. (1996). Vocational Education in the United States: The Early 1990s. Washington, DC: US Department of Education.

National Council of Teachers of Mathematics. (1989). Curriculum and Evaluation Standards for School Mathematics. Reston, VA: Author.

Packer, A. (1997). Mathematical Competencies that Employers Expect. In L. Steen (Ed.), Why Numbers Count: Quantitative Literacy for Tomorrow's America (pp. 137-154). New York, NY: The College Board.

Ravitch, D. (1995). National Standards in American Education: A Citizen's Guide. Washington, DC: The Brookings Institution.

Secretary's Commission on Achieving Necessary Skills (SCANS). (1991). What Work Requires of Schools: A SCANS Report for America 2000. Washington, DC: US Department of Labor.

Shulman, L. F. (1997). Professing the Liberal Arts. In R. Orill (Ed.), Re-imagining Liberal Learning in America (pp. 151-173). New York: The College Board.

Wadsworth, D. (1997). Civic Numeracy: Does the Public Care? In L. Steen (Ed.), Why Numbers Count: Quantitative Literacy for Tomorrow's America (pp. 11-22). New York, NY: The College Board.

CHAPTER 9

MATHEMATICS KNOWLEDGE AS A VOCATIONAL QUALIFICATION

TINE WEDEGE

Abstract. The concept of qualification offers a framework for didactic reflection on the relation between education and work. It is a general view among educational planners that mathematics is relevant in the context of providing qualifications for the labour market. Mathematics knowledge does not, however, become a vocational qualification unless it is integrated with knowledge, skills and properties that are relevant in relation to technique and organization in the workplace. Furthermore, adults' need of math-containing qualifications must be perceived from both a subjective and an objective point of view, i.e. from the point of view of individual workers and not only from the point of view of the labour market.

1. INTRODUCTION

Adult education programmes are highly prioritised in educational and labour-market policy in the European Union, primarily with reference to technological development. Educational planners are of the opinion that they can find objective grounds for objectives and priorities in the qualification analyses which began to appear in earnest during the 1980s. Furthermore the concept of qualification[1] is an important link between the social and pedagogical research fields where studies in the subject area of 'adult education for mathematics in the workplace' are situated. The concept provides a framework for didactic reflection on the relation between adult education and work. The two traditional research questions are: What are the qualifications needed for technological development on the labour market, and how can they be translated into education (or qualification)?

In policy reports the general categories of qualifications are described in isolation from the technological contexts of workplaces. Mathematical knowledge as such, referring to the formal disciplines (arithmetic, algebra and geometry), is pointed to as a key competence. (Darrah, 1992). In addition, educational planners most often use a simple equation:

> qualification demand - qualifications = qualification need
> when translating the qualification demand of the labour market to workers' need of qualification. (Wedege, 1995)

As a result of this way of thinking, mathematics is centrally placed in adult education programmes. I shall argue here that we must ask the following question:

Bessot & Ridgway (eds.), Education for Mathematics in the Workplace, 127—136.

How does mathematics knowledge integrate into vocational qualifications? Furthermore, I shall briefly discuss how mathematics is hidden in modern technology and argue that adults' experiences with mathematics should be seen in the duality between objective demands of the labour market and subjective embedding, i.e. the adults' motives for learning, or not learning, mathematics.

The *'adults'* that I am speaking of are those with brief schooling whose perspective with regard to education is about training themselves for the job on hand or for skilled work. My interest is that of an educational planner, but my point of departure as a researcher is that reforms in adult education must address the social nature of work, including the capacity to understand and modify technology. (Wedege, 1995 a & b).

2. SOME BASIC CONCEPTS

I shall define *technology* on the labour market as consisting of three elements - technique, human qualifications, and work organization - and their dynamic interrelation. *Technique* is used in the broader sense to include not only tools, machines and technical equipment, but also cultural techniques (such as language and time), and techniques for deliberate structuring of the working process (as for instance in Taylor's 'scientific management' and ISO 9000 quality certification). *Work organization* is used to designate the way in which tasks, functions, responsibility, and competence are structured in the workplace.

I define human *qualifications* as the knowledge, skills and properties that are relevant to technique and work organization as well as to their interaction in a work function. I distinguish analytically between three types of qualification:

-*specific professional qualifications*: technical-professional knowledge and skills that are directly and visibly present when the individual work function is being carried out;

-*general qualifications*: general and professional knowledge and abilities such as literacy and numeracy that are (often indirectly) present when a wider range of work functions are being carried out; and,

-*social qualifications*: personal traits/attitudes that are present in the work process such as precision, solidarity, flexibility and ability to cooperate. (Wedege, 1995 a & b)

These three types of qualifications are interwoven in the single individual. A skill or understanding might be analysed as a specific professional qualification in a work context and as a general qualification in another context. For example, skill in reading diagrams and applying this knowledge is a specific qualification for the driver of a fork-lift truck, while skill in reading and understanding a chart of absence due to sickness is a general qualification for workers.

Human qualifications constitute a central element in technology where they are utilised and developed in interaction with technique and work organization. On the basis of this conception of technology and technological development it is necessary to distinguish between necessary and relevant qualifications in analyses which are to be used for purposes of educational planning. For example, a given technique with different methods of organizing work may require different types of qualifications. (Wedege, 1995a)

3. MATHEMATICS IN THE DANISH ADULT EDUCATION SYSTEM

Concerns about people who are ill-prepared for the workplaces of the future have resulted in various policy reports identifying 'basic skills' and 'key competences' and postulating appropriate educational responses. As one of the basic skills, 'mathematics' is the most widely-used general subject in the educational programmes available. In Denmark more than 100,000 unskilled and semi-skilled workers participate every year in mathematics education or training and even more in math-containing further education.

By *mathematics instruction* I mean organized communication of a mathematical subject area, either as single-subject teaching (in a formal or informal context) or as part of an educational programme as an independent subject or module. By *math-containing instruction* I mean organized communication of a single or interdisciplinary subject area where mathematics is an integrated but identifiable part. The instruction can be informal (for example, learning from one's colleagues at a place of work) or be part of a course or a study programme.

In Denmark, the Adult Education System offers a wide-ranging set of learning opportunities to adult early school leavers. Each of these has its own legislative foundation, regulating contents, organization, and financing, as well as its own institutions. The objectives of the different adult and further education programmes range from a purely job or training perspective, over a broad vocational to a societal perspective. The two biggest national programmes are Adult Vocational Training (AVT) and Formal Adult Education (FAE) which are to be found in two different sectors.

-Under the Ministry of Labour: AVT. Courses providing unskilled and semi-skilled workers, skilled workers and middle management with formal qualifications. The courses are aimed at work functions in a great number of industries (building and construction, the commercial and clerical area, the metalwork industry, transport etc.). Adult Vocational Training is organized as alternance training with periods of work between the courses. Much of the instruction takes place in practical training premises that resemble work places.

-Under the Ministry of Education: FAE - general adult education on lower secondary level consisting of single subjects such as Danish, Mathematics, History, Physics, English etc. Passing an examination in a subject gives the same formal competence as the leaving examination from the Folkeskolen (primary and lower secondary school) but both teaching and examinations are specially designed for adults.

Mathematics instruction can be found in both sectors but the point of departure for both curriculum and objectives is different. In Formal Adult Education the starting point is mathematics as system and method and the objective is knowledge and abilities concerning mathematics as a subject in itself and in relation to its uses in everyday life. The teachers are often primary and secondary school teachers who have received supplementary training in how to teach adults. In Adult Vocational Training the curriculum is based on an assessment of the need for mathematics knowledge in the different job functions, and the objective is knowledge and abilities concerning mathematics in relation to other subjects. The teachers are skilled workers or technicians with medium-cycle education who have taken a supplementary course in pedagogics. Although mathematics instruction as such does not play any significant part in AVT, math-containing instruction, either as theory (e.g. understanding a working drawing) or as practice occupies approximately 1/5 of the timetable on average. The teaching objectives are action-oriented and formulated as providing competence to perform certain work functions.

4. TWO DIFFERENT CONCEPTIONS OF MATHEMATICS KNOWLEDGE

In recent years these two adult educational systems (Adult Vocational Training and the Formal Adult Education) have been cooperating in cross-sectorial courses to provide adult early school leavers with vocational qualifications. When the systems cooperate, mathematics is the most widely used general subject.

As an example of a clash between two different conceptions of mathematics knowledge as a qualification, I have chosen a case from a cross-sectorial course where a FAE mathematics teacher and an AVT specialist teacher are teaching the same group. The objective is to ensure that the participants in one and the same course achieve a formal general competence in mathematics and professional competence in the wood industry:

> During the preceding week, the participants have been working in the wood workshop. They have marked and sawed out a tabletop shaped like an equilateral octagon. The teaching of mathematics is therefore concerned with octagons, including the isosceles triangles they are composed of.
>
> We join the class at the point where the teacher, having reached agreement with the participants on the 'diameter', is now preparing to calculate the size of each edge - that is the base of the equilateral triangles.
>
> One of the participants spontaneously leaves the room with a remark to the effect that she is going down to measure how long they are.
>
> The teacher continues to explain how the base line can be measured. In the middle of all of this, the participant returns.
>
> The teacher breaks off, asking: *Well, how long was it then?*
>
> The participant — *It was 43.3, that is one of them ... the other was 44.2!*
>
> The teacher: — *But it can't be right! They ought to be the same size.*
>
> Participant, quickly: — *But it's true.*
>
> *Alright, if you have measured it we'll write it,* answers the teacher obligingly.
>
> Whereupon the participant spontaneously retorts: *Here we go again! You've got your theories; it's different in practice.* And there is scattered applause and remarks by the audience. (Scavenius & Wahlgren, 1994, pp. 132-33)

In the world of this mathematics teacher it was unthinkable that the edges of an equilateral octagon were not of exactly the same length. In this particular wood workshop an equilateral octagon was an octagonal polygon whose edges deviate by a maximum of 1 cm. In the world of work/realities this is called the 'tolerance', and it varies from trade to trade and from product to product.

5. MATHEMATICS KNOWLEDGE AS A QUALIFICATION

Decisions to upgrade adult education and to place mathematics as a central subject are based on two assumptions. Firstly, that education is an appropriate answer to adults' need of qualification, and secondly that mathematics knowledge is a relevant vocational qualification. It is, in fact, a general view among Danish educational planners that mathematics is relevant as a general qualification in the context of providing professional qualifications. Mathematics knowledge does not, however, become a qualification as defined above unless it is integrated with knowledge, skills and properties that are relevant in relation to technique and work organization and in the interaction that takes place in a work function. As Rudolph Strässer concludes a detailed study of thirty metalwork draughtsmen:

> Interpreted globally, the study clearly shows that mathematics and vocational knowledge are intimately interwoven at work. Workplace practices do not distinguish mathematical knowledge from other knowledge helpful to cope with the professional problem. (Strässer, 1996, p.5)

In order to serve as a qualification, mathematics knowledge must be able to prove itself in the work process. I contend that mathematics knowledge as a qualification inevitably includes practical and/or reflective (mathematics) knowledge.[2] This claim is based on the thesis that *just because you know mathematics you do not automatically know how to use it*. That is why in the definition of general qualification I use 'numeracy' to mean functional mathematics skills and understandings, not 'mathematical knowledge' as such.

> [...] mathematics in practice is more unpredictable and far less pretty than is the landscape of logic and deduction. It is rich in data, interspersed with conjecture, dependent on technology, and tied to useful applications. (Forman & Steen, 1995, p. 227)

In order to identify and describe mathematics in semi-skilled job functions and to analyse how mathematics knowledge at work is interwoven with specific qualifications and social qualifications, I am investigating selected firms within four lines of industry: building and construction, the commercial and clerical area, the metal industry, and transport. In organizing my investigation I am following the systematics developed in the Australian project, 'Rich Interpretation of Mathematical Ideas and Techniques'. I shadow a worker for half a day to describe the action taking place. At the end of the day I interview the worker to explore any issues that have arisen. The observations are written up as a descriptive story with examples of particularly interesting incidents. (Hogan, 1997; see also Hogan in this section.) Furthermore, I photograph interesting situations and tools and collect written materials with numberss, formulas, diagrams etc. (such as working drawings, plans and statistics). In processing the data I use a working model with four analytical dimensions which is constructed to describe and analyse numeracy. One is *media*; the relevant numeracy depends on whether it is to be applied to oral information, a manual, the week production plan, or a pile of earth, even if the figures and the four basic arithmetical operations are the same. *Context* is another dimension; what one knows and what one should know depends on whether one is in a supermarket, at work, or in a test situation. In my investigation the context is described in terms of work function and technology. A third dimension comprises *personal intention*; it is crucial whether one wishes to obtain or to give information, to plan production, to check the quality of the product, to pass the time etc. The fourth is *skills and understanding* such as geometric sense, rough estimates, competence in mathematical modelling and a sense of sizes. (Lindenskov, 1997; Wedege, 1998)

In the case of a semi-skilled worker the ability to add and subtract is a general qualification, that is, if she can do it in connection with solving a specific professional problem. Her mathematical knowledge about addition and subtraction algorithms must be available for use outside the classroom, where in a concrete situation she must also be able to mathematicize a problem and decide whether the numbers must be added or subtracted (practical knowledge).

Example 1. A semi-skilled worker in a large metalwork factory is turning shields for pumps on a CNC machine. She checks each item individually. The measurements are within the limits of tolerance, but she is not personally satisfied with the quality: there is a dark ring left at the bottom of the item she has turned. The worker adjusts the machine (the tool has to be moved further in to correct this fault) by trying out different adjustments: first by 5/10 and then by 2/10. She does this by subtracting 0.5 and 0.2 on the screen and she says that it is always necessary to consider whether to add or subtract.

In this case the tolerance is +/- 0.1 mm. In a working situation reconciliation of units such as m, cm, or mm will always be involved (practical knowledge). In a problem-solving situation reflective knowledge may be involved:

Example 2. At a large electronics factory a semi-skilled worker with many years of experience in production is now working in the department where blanks from a subcontractor are subjected to quality control. She takes a bag containing 9 small connectors for flat cables which must be checked in relation to various standards. It is a new brand and one of the measurements, taken with a digital slide gauge, does not quite fit in with the drawing. It is 15.58 mm and should be 16.00 mm according to the drawing. The tolerance is 0.01, but the worker's experience from the production department now benefits her - and the factory. She told me that the discrepancy in this measurement and this connector has no practical significance. Had the measurement been over 16.01 she would have rejected the items.

Her general knowledge about reading and understanding workshop drawings should be applicable in a specific professional qualification which comprises knowledge about using the items which have been checked. In practice this means that she does not merely reject items because they do not meet the requirements but she uses critical judgement in the situation in her translation of the specifications. She can also see on her computer screen that during the day the production department will be short of this type of connector.

Up to now my investigations of mathematics in semi-skilled job functions supports Forman and Steen's evaluation of the need for concrete mathematics in vocational education:

> Concrete mathematics, built on advanced applications of elementary mathematics rather than on elementary applications of advanced mathematics. (Forman & Steen, 1995, p.228; see also Forman & Steen in this section)

On the basis of this brief discussion of mathematics knowledge as a qualification, I should like to conclude as follows. Teaching of mathematics aiming at qualifying the participants for the labour market and, consequently, development of practical and reflective knowledge, cannot be based on traditional conceptions of subjects and demarcations. It must reach across established subject demarcations, precisely because the reasons for choice of material and content are derived from outside, not from within the subject of mathematics.

6. THE OBJECTIVE AND SUBJECTIVE PERSPECTIVE

In my definition of the concept of qualifications, I speak of relevant knowledge, skills and properties rather than of necessary knowledge etc. This makes it possible to perceive qualifications from two different perspectives: subjective and objective, i.e. from the point of view of individual workers as well as from the point of view of the labour market. As in the example above from the cross-sectorial course, participants in adult vocational training are often surprised at or sceptical about the fact that mathematics is a subject of study in their programme. Their learning perspective is vocational qualification, not to learn mathematics. (Strässer & Zevenberger, 1996; Wedege, 1999) But mathematics is interwoven in technology -in technique, work organization and qualifications. However, modern computer technique hides the use of mathematics in the software, and mathematics as a visible tool disappears in many workplace routines.

Example 3. At Copenhagen airport different function groups cooperate by means of computer. When the aircraft are being loaded and unloaded, the loading group and the load planner are in constant computer contact. In the loading instructions the planner has placed baggage, cargo and mail in the four cargo compartments in front of and behind the wings. The ideal balance factor (38.0) and the limits (5.9/51.6) also appear from these instructions. The balance factor of the aircraft can also be read on the screen during loading: In the loading report for the specific aircraft it is 28.2. The work team does not have to enter the figures for the distribution of weight between the four cargo compartments in the formula for the balance factor as this figure is automatically calculated when the cargo and the weight are entered.

When decisions are being made at the airport about loading an aircraft, the priorities are 1) safety, 2) keeping to the timetable, and 3) service. Time is often at a premium when an aircraft is to be unloaded and loaded, and keeping to the timetable can mean that some cargo may have to be sent on a later plane. As the first priority is safety, this can, however, mean that the flight may be delayed and the level of service may not be so high if the balance factor is not within the permitted limits. The authority to release the flight belongs to the foreman of the group but the decision not to follow the loading instructions can only be made after consultation with the loading planner.

This is an example of the fact that the necessary qualifications are not only to be found in technique alone. The balance factor for the load in question can be read on the computer. This means that the foreman of the group, without any knowledge of the formula for the balance factor, has a means of measuring what happens if they leave some cargo out of the loading instructions. In principle he can make this type of decision but if there is extra cargo which does not figure in the instructions (and this was actually the case with a later plane), he cannot precisely predict what will happen to the balance factor if the cargo is placed, for example, in compartment 4. If the work were organized in autonomous groups, knowledge of calculating the balance factor would have been necessary to carry out the job function.

Calculating the balance factor is actually part of the courses of training for this work. All the companies where I make observations use Adult Vocational Training to qualify their workers. However, the invisibility of mathematics in work (Harris, 1994) combined with the adults' learning perspective result in missing motivation to learn mathematics. Mogens Niss has called this phenomenon the contradiction between the objective social relevance of mathematics and its subjective irrelevance, 'the relevance paradox'. (Niss, 1994)

The current demand for qualifications on the labour market has been investigated in a number of areas and occupations, and educational measures have been taken accordingly. But this kind of objective perspective on qualifications is insufficient as it does not take into account the fact that qualifications are integrated characteristics of human beings and cannot be 'produced' or applied without being influenced by subjectivity.

> Qualifications must be understood in their duality between the objective demands that determine them and the subjective imbedding that constitutes their conditions of existence. (Andersen et al., 1996:28)

This is the most central conclusion in the 'General Qualification Project', a large-scale research project on the development of general qualifications in Danish adult education. (Andersen et al. 1996; Olesen 1996) They have developed a model for 'subjectivity in the perspective of qualification' which is suited to maintaining the double subjective and objective perspectives as the basis for understanding qualification processes as they take place in programmes of education, at work and in other life contexts.

In this model, qualifications are distributed on three levels of subjectivity: a basic level, a comprehensive level and a specific level, indicating three different ways in which qualifications are linked to the subject's self-perception and the strength of the respective links. With inspiration from this search model, on the basis of a questionnaire survey (teachers' impression) and participant survey (participants' experience) conducted at the Adult Vocational Training Centres in 1995-96 (Lindenskov, 1996; Wedege, 1997), we can analyse participants' experience with mathematics on three levels (figure 1).

The level of skills	Specific skills in arithmetic and mathematics which are a visible part of the work process.
The level of understanding	General mathematics knowledge, e.g. understanding and dealing with the theory-practice relation in the working situation.
The level of identity	A judicious mixture of incorporated skills and understanding (mathematics thinking, tacit knowledge) *and* attitudes, feelings and motives.

Figure 1. Three levels in participants' experience with mathematics

I suggest that we look at the three different types of mathematics knowledge (mathematical, practical, reflective) through the lens of these three levels, and we may see their interrelationship in adult's experience with mathematics. Mathematical knowledge as such is embedded at the two levels of skills and understanding; practical (mathematics) knowledge is embedded at all three levels, while we find reflective (mathematics) knowledge at the level of understanding.

As a provisional conclusion to these considerations about technology, adults, and mathematics, I should like to conclude with the following statements about mathematics teaching in adult vocational training and education:

The mathematics teaching of adults has to take the participants' subjectivity into account, meaning adults' experience at all three levels of subjectivity including their perspective for learning, or not learning, mathematics. The relevance paradox can be thematised in teaching by making explicit the mathematics in technology at the workplace: technique, work organization, and qualifications.

The research work is partly funded by the Danish Research Councils.

7. REFERENCES

Anderson, V., et al. (1996). *General Qualifications. 9th Report from the General Qualification Project.* Roskilde: Adult Education Research Group, Roskilde University.

Darrah, C. N. (1992). Workplace Skills in Context. *Human Organization, 51*(3), 264-273.

Forman, S. L., and Steen, L. A. (1995). Mathematics for Work and Life. In I.M. Carl (Ed.), *Prospects for School Mathematics: Seventy Five Years of Progress* (pp. 219-241). Reston, VA: National Council of Teachers of Mathematics.

Harris, M. (1994). *Finding Common Threads: Researching the Mathematics in Traditionally Female Work.* Paper presented at the Adults Learning Maths - A Research Forum. ALM-1. Proceedings of the Inaugural Conference, London.

Hogan, J. (1997). *Rich Interpretation of Using Mathematical Ideas and Techniques (RIUMIT). Final Report to the Commonwealth Department of Employment, Education, Training and Youth Affairs.* Adelaide: Australian Association of Mathematics Teachers Inc.

Lindenskov, L. (1996). *"Det er Fordi Jeg Mangler Billeder..." AMU-kursisters Oplevelser og Potentialer i Faglig Regning og Matematik.* Copenhagen: Directorate General for Employment, Placement and Vocational Training.

Niss, M. (1994). Mathematics in Society. In R. Biehler, et al., (Eds.), *Didactics of Mathematics as a Scientific Discipline* (pp. 367-378). Dordrecht: Kluwer.

Olesen, H. S. (1996). A New Concept of Qualification. In H. S. Olesen, P. Rasmussen, (Eds.), *Theoretical Issues in Adult Education - Danish Research and Experiences.* Roskilde: Roskilde University Press.

Scavenius, C., and Wahlgren, B. (1994). *Tvaersektoriel Undervisning. Femte og Afsluttende Rapport om Samarbejdet Mellem AMU og VUC.* Copenhagen: Research Center for Adult Education, The Royal Danish School of Educational Studies.

Skovsmose, O. (1990). Mathematical Education and Democracy. *Educational Studies in Mathematics, 21*, 109-128.

Straesser, R. (1996). *Mathematics for Work - a Didactical Perspective (Regular lecture).* Paper presented at ICME 8, Sevilla.

Sträesser, R., and Zevenbergen, R. (1996). Further Mathematics Education. In A. Bishop, K. Clements, C. Keitel, J. Kilpatrick and C. Laborde, (Eds.), *Mathematics International Handbook on Mathematics Education* (pp. 647-674). Dodrecht: Kluwer.

Wedege, T. (1995a) Teknologi, Kvalifikationer og Matematik. *Nordic Studies in Mathematics Education, 3(2)*, 29-52.

Wedege, T. (1995b). *Technological Competence and Mathematics.* Paper presented at the Mathematics with a Human Face Conference. Proceedings of Second International Conference of Adults Learning Mathematics, London.

Wedege, T. (1997). *Profile in Mathematics of Adults Returning to Education.* Paper presented at Working Group 18, ICME 8, Adelaide.

Wedege, T. (1998). Adults Knowing and Learning Mathematics - Introduction to a New Field of Research between Adult Education and Mathematics Education. In S. Tosse, et al., (Eds.), *Corporate and Nonformal Learning. Adult Education Research in Nordic Countries* (pp. 177-197). Trondheim: Tapir Forlag.

Wedege, T. (1999). To Know or Not to Know - Mathematics, That is a Question of Context. *Educational Studies in Mathematics, 39*(1-3), 205-227.

8. NOTES

[1] The English term 'qualification' covers both actual qualification(s) and the process of becoming qualified. I use 'qualification' in both senses.

[2] I use Ole Skovsmose's distinction between three types of mathematics knowledge:

1. Mathematical knowledge as such.
2. Technological knowledge, which in this context is knowledge about how to build and how to use a mathematical model.
3. Reflective knowledge, to be interpreted as a more general conceptual framework, or meta-knowledge, for discussing the nature of models and the criteria used in their constructions, applications and evaluations. (Skovsmose 1990:124)

For pragmatic reasons, I prefere the term 'practical knowledge' instead of 'technological knowledge'.

SECTION 3

EDUCATING FUTURE WORKERS

SECTION 3

EDUCATING FUTURE WORKERS

PREFACE

JULIAN WILLIAMS

The contributions to this section all address the problem of the development of a mathematics curriculum which is sensitive to the vocational aspirations and needs of future workers. The problem at issue is one that has been heralded by the previous two sections:

-Mathematical knowledge constructed in school is not straightforwardly 'transferred' to outside school practices.

-Mathematical practice in the workplace often seems to have different characteristics from that in College.

-For many students, mathematics in school and College lacks a sense of authenticity.

In each chapter these issues are again touched on, but only indirectly in so far as they help us to understand how to better develop the mathematics curriculum. After all, we cannot wait for these big problems to be solved and their solutions applied to educational practice. Indeed the reshaping of education for future workers may be part of the process of coming to understand these problems better.

Annie Bessot examines the geometrical knowledge underpinning diagrams and plans for the building site. In particular she is interested in how problematic drawings are constructed by students geometrically. She developed a special experimental programme to try to help students overcome well-known difficulties with visualising and drawing of plans, making use of some innovative tools and authentic problems. It seems that geometrical knowledge is developed rather than applied in the solution of the problems posed, and this leads her to conclude that 'geometrical knowledge is working knowledge for developing solutions to problematical tasks'. Thus, the curriculum might be reshaped and a new synthesis of 'geometry' and 'working knowledge' may resolve the three problems listed above.

A similar problem is tackled by Corinne Hahn: jewellery apprentices typically confront inauthentic problems and procedures in College mathematics. Certain

Bessot & Ridgway (eds.), Education for Mathematics in the Workplace, 139—141.

percentage calculations taught in College, if used in the workplace, would lead the employer to be sued for malpractice! She further observes that the use of realistic, 'contextualised' problems in the College may do little to help: she poses instead professional situations, which authentically model the situations arising in the workplace. In such situations it seems that students are more likely to build appropriate solutions, and less likely to make common conceptual errors. Apparently the apprentices professional 'sense' supported flexible mathematical constructions.

In both cases, the authors mention that their work represents a departure from the normal authorised curriculum. Moving from experimental programmes to large-scale reform is notoriously difficult. In the third chapter, Geoff Wake and I describe a prevocational curriculum and assessment system developed in the UK intended to facilitate such innovations. The design is based on the concept of 'general mathematical competence', which attempts to synthesise the particularity of competence with the generality of mathematics. In this scheme the curriculum lays down the mathematics to be learnt, but the assessment system allows that the mathematical competence be demonstrated in practice through portfolios of practical work.

All three chapters leave us with unresolved problems and open questions; and since they were written no doubt all three programmes have developed their research further. All three chapters suggest that the achievement of an appropriate 'synthesis' of College mathematics and practices of the workplace is problematic. I would like to highlight three practical questions arising implicitly from my reading of the chapters that seem worth making explicit:

-Are the syntheses of College mathematics and workplace practice we propose really appropriate to the *future* world of work?

-What teacher-knowledge is implied by the reforms we put forward, and what are the consequences of these for reform?

-Are the educational systems in which we work really compatible with reforms aiming to make mathematics learning authentic?

Research in the field can be seen historically. In the early days of vocational education researchers asked employers what mathematics they needed, and used these to construct tests and syllabuses. Then more critical research observed workers and described what mathematics knowledge workers used. This paradigm suggests changes in curriculum without necessitating major changes in practice. In the modern phase mathematically-minded anthropologists looked at practices in workplaces and on the streets and contrasted these with school mathematics practices. Seeing these practices as socially situated pushes curriculum developers to design new social arenas and systems for mathematical activity that might create some space for students to learn authentically.

The chapters in this section describe some attempts to do this, but the questions they pose suggest just how challenging this task might be. How can schools prepare for future workplace practice when we know so little about it, and *can* only know so little about it, in the present? Since schooling is a social practice, how can we expect teachers whose lives are thereby circumscribed to help students to break out of it? Finally, school systems themselves have many conservative institutional features based on powerful vested interests, and it is not clear where challenges to schooling practices will find the social forces required to implement them. Almost certainly the solution of these problems must involve a further breakdown in the division in practice between schools and workplaces, teachers and workplace trainers and learners and workers. The chapters in this section reflect only the beginnings of a move in this direction.

CHAPTER 10

GEOMETRY AT WORK

EXAMPLES FROM THE BUILDING INDUSTRY

ANNIE BESSOT[1]

Abstract. What role does geometrical knowledge play in solving spatial problems in both vocational training and the workplace? Two activities are fundamental to the solution of spatial problems: measurement and the graphic representation of objects in space. Our 'didactical engineering' attempts both to introduce technical drawings as a geometrical model (in relation with orthographic projection), and to simulate some of the practicalities of reading plans on the building-site: the conditions necessary for this to happen are created by certain parameters of a basic reading situation. The teaching scenario, set up here, facilitates the transfer to students of part of the responsibility for using sophisticated geometrical knowledge in order to develop techniques for marking out.

1. INTRODUCTION

Before it becomes scientific knowledge, elementary geometry is an effective model for the practical treatment of problems which involve shape and space such as those experienced by professionals in the building industry. The Greeks, for reasons in which cultural factors seem to play a decisive role, were the inventors of 'mathematical' geometry and at the same time the inventors of the notion of mathematical proof (Arsac, 1987). Geometry since that time has developed more and more as autonomous theory, to the point where it often appears to have broken with its spatial origins: the geometrical properties which are of interest in 'mathematics' and those concerned with solving problems in space are not the same. The resolution of apparently similar problems is motivated by different concerns: the mathematician seeks truth whilst the practitioner looks for maximum efficiency. A builder may well have problems finding solutions to practical problems in a traditional geometry course.

> the use of geometry as a technology of space has looked, since Classical times, like a set of *culturally governed*[2] social practices (Chevallard 1991, p. 60)

Because of this, the application of geometrical knowledge to training for jobs in the building industry turns out to be a difficult problem for teaching. In France, it has occurred through the appearance of descriptive geometry (which we owe to the mathematician Monge), and its manifestations (practical geometry), followed by the

Bessot & Ridgway (eds.), Education for Mathematics in the Workplace, 143—157.

virtual disappearance from vocational training of geometry as the structured entity commonly found in school mathematics.

2. GEOMETRY AND THE READING OF PLANS ON THE BUILDING-SITE

Two main trends have informed attempts to control space: measurement and graphic representation of objects in space[3]. Plans and technical drawings are 'tools'[4] which allow the spatial environment to be manipulated, but they are also mediators which are socially constructed and passed on in order to resolve the contractual problems of communication between designer, client and those carrying out the work.

Unlike the drawings which occur in elementary geometry, a plan is *a working drawing* in which *specific measurements* play a central role. Measurements may or may not be given by markings on the drawing.

It is the linear invariants of orthographic projection which provide the theoretical basis for understanding relationships between *measurements*. The reading of plans is partly connected to *geometrical knowledge* but cannot simply be reduced to this. One of our surveys (Bessot & al., 1992) on the practice of reading plans on building sites bears out the fact that three situations (not necessarily following on from one another) condition different practicalities of reading:

-The worker[5] receives one or several plans from the project manager: he must first of all determine where the work is to be carried out, and for this it is necessary to link up

> two spatial systems of reference implied by the (reciprocal) transfer of a real space to its representation on a plan (Weill-Fassina, Rachedi, 1993).

A plan then functions as a tool *which is comparable to a map*, allowing the worker to move about and locate areas within the space of the building site.

-Once the location of the work has been ascertained, the purpose of reading the plans changes. It is at this moment that the plans become *tools for actually carrying out the work.* The profession distinguishes at least two tasks linked to the reading of a technical drawing: *Setting out*[6] and *Marking out*[7]

-Finally, a plan may be, on the one hand a *means of checking* the intermediate stages of the work by the worker and the project manager, and on the other *a means of anticipating and foreseeing and/or managing* any conflicts which may arise from the necessary interactions between the different parties involved: for example, tolerances accepted in the work of builders must still allows carpenters to put in the door casings.

3. BETWEEN GEOMETRY AND SPACE: THE DRAWING AS A BASIS FOR MARKING OUT THE BUILDING LINES

Let us consider, amongst the building site practices, the practice of reading a work plan in order to *mark out the site* of the building on the ground.

> The markings are [...] a full scale representation of those parts of the work to be carried out from the drawings made, either by the architect (for example, a plan of the

> arrangement of partition walls), or by the design office (for example, a plan to case a floor in reinforced concrete) (Olivier[8] 1976, p. 19)

Drawings as well as markings have the characteristics of *full-scale working drawings* which require equivalences to be established between *measurements*. These measurements and the checking of them lead on to considerations of scale (when dimensions are absent), precision, tolerance and quality. The drawing or plan thus has two facets, with one side directed towards real space and the other towards geometrical space. On a plan, a window is represented by the drawing of a rectangle, and a beam is shown by a line segment. In reality, there are no lines which are mathematically straight, nor surfaces which are 'flat', and therefore no rectangles nor line segments: the drawing is a model both of the beam and of the ideal object represented by the line segment.

The plan given to the worker is assumed to contain sufficient information to mark out a geometrical contour on a 'flat' surface. However the problems presented by marking out are very different from the problems of mathematical construction which concern 'the possibility or impossibility of a geometrical solution.' (Klein 1896). The solution to a problem of mathematical construction must include mathematical proof of the existence of the 'object' to be constructed and of its constructability; drawings, which are physical representations of 'ideal figures' of the same name, may possibly be included in order to give a 'graphic experience' of the study being carried out. Professionally, the process of marking out is feasible as long as it provides a full-scale working drawing which is sufficiently precise in relation to the norms of tolerance, the primary concern being effectiveness. Therefore it is possible to find a number of processes[9] which may be used by builders to construct a partition perpendicular to a wall; one such process involves the use of 'the 3-4-5 rule'. From a 'point' A, marked on an existing wall and represented by a line on the plan, the builder traces *on the ground* an ABC right-angled triangle whose sides measure 3-4-5 metres[10], with the help of a two-metre or a ten-metre ruler. Familiarity with the 3-4-5 rule allows the processes of the work to be carried out successfully, but knowledge of the elementary geometry which explains these processes is entirely unnecessary for the worker.

However, for practitioners and those involved in training, techniques for marking out derive from geometrical knowledge which needs to be taught:

> most of the procedures used on building sites come from geometry, and it is therefore necessary to have some basic knowledge of this science (Olivier 1976, p. 19)

In some *non-mathematical* textbooks currently used in vocational training, practical procedures are used to justify the geometrical properties taught. For example, to quote Olivier again:

> *The 3-4-5 rule*
> For more than 2500 years, builders have been using the 3-4-5 rule, very often without knowing the equation which links these three numbers, but the connection is easy to check: (3x3)+(4x4)=(5x5), which is stated as follows: *'The square built on the hypotenuse of a right-angled triangle is equivalent to the sum of the squares built on the sides of the right angle'* [...]. (ibid.pp. 42-43)

For those in charge of vocational teaching for construction work the need to have geometrical knowledge expressed in this way appears to result from the necessity of explaining why a particular method or process is used; it is a sort of 'ecological' law of science teaching, in the sense that knowledge is organic, continually subject to change and modification. Chevallard (1996) expresses it as follows:

> [If] one calls the *technology* of a technique τ a 'discourse' which justifies τ [...], in any institution I, the existence of a technology, even a crude one, is an essential ecological condition for the existence of a technique. (op.cit. p. 17)

Closely connected to this need for justification when transferring practical skills, is a concern for effectiveness which also appears to underlie the teaching of geometry. If the task becomes problematic, that is to say if the boundaries of *customary* working practice are crossed, it can be expected that the worker will have to rely on explicit geometrical knowledge in order to find a practical solution[11].

The next section considers the role of geometrical knowledge in the linking of drawings and objects in space.

4. LINKING DRAWINGS WITH OBJECTS IN SPACE: TEACHING STRATEGIES

In France, from Monge's descriptive geometry (1796) until the present day, Deforges (1981)[12] notes two opposite strategies for teaching technical drawing: namely orthographic views, and systems of orthographic views. These methods, direct and indirect, are differentiated according to the knowledge bases chosen to justify them.

Direct method

> The most direct justification for geometrical representation of the structure of a building is that it is the drawing of what an observer might see looking full square at the building with one eye, in such a way that the gaze is always perpendicular to the side observed. (op.cit.)

The training procedures based on this method assume, on the one hand that the activity of perception reflects an area of knowledge which allows the person to organise the relationship between drawing and actual object in such a way as to be able to act when faced with reading the drawing, and on the other hand that this knowledge can be learnt.

Indirect method

A view of an object in space is no longer what is perceived by looking at the object; it is a drawing similar to that which can be obtained through orthographic transfer of particular points, linked to the geometrical model of the object, on a reference plan: the notion of orthographic projection justifies the relationship between the actual object and the drawing. To quote Monge's comments (1796) on descriptive geometry:

> The first objective is to represent precisely, in drawings which have only two dimensions, objects which have three, and which can be strictly defined[13]. [...] The second objective of descriptive geometry is to deduce from the precise description of solid objects all that this necessarily implies about their shapes and respective positions. [...] it is indispensable for all workmen whose aim is to give an object a certain pre-determined form. (op.cit.p.2)

Figure 1 which follows shows the complexity of the links set out in this geometrical modelling[14].

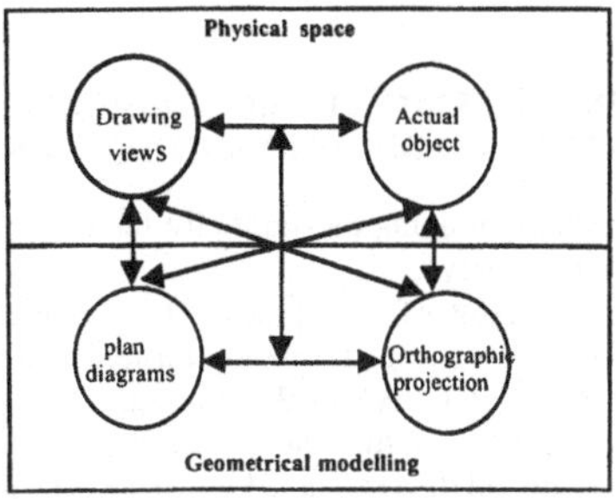

Figure 1. Relations between an object in space, a drawing and a geometrical model

The training procedures based on the 'indirect method' pre-suppose that knowledge of the geometrical relationships between orthographic projections and plan diagrams will allow decisions to be made when reading them, and will allow the user to check the relationship between the drawing and the actual object to be created.

The dichotomy between direct and indirect method introduced by Deforge is useful because it highlights the different forms of knowledge favoured in the discourses which are used to justify the techniques taught (drawing[15] as well as reading). In the case of the direct method, the first checks made to validate a piece of work produced are based on visual perception: the invariants of the relationship between object and drawing can thus only be the result of observations based on local perception which must be generalised into signs and symbols on the plan. The direct method, being based on perceptual judgement, therefore saves expressing the notion of view mathematically.

On the other hand, the indirect method, by rejecting the role of perception, poses a real challenge. The complexity of the relationship between the object and its orthographic projection increases the risk of the observer falling back into a way of seeing which is controlled by his own perception. Deforge shows that the official existence of other methods of justification for the notion of view results from the difficulties encountered by students when faced with the indirect method in basic vocational training.

Although more than half of the present-day educational French textbooks that we consulted justify the introduction of views by the indirect method (10 textbooks out of 19 studied), they very quickly abandon any reference to orthographic projection when it comes to surveying, which is the focus of students' work. The assumption can be made that habits then take root in the practices of reading and drawing which cannot be adapted when small variations in tasks occur. In France there is a large bibliography on the difficulties observed in these practices, compiled in particular by researchers into the psychology of work (see for example Rabardel, Weill-Fassina, 1987).

5. DIDACTICAL ENGINEERING[16]: BASIC TRAINING PROCEDURES FOR THE READING OF TECHNICAL DRAWINGS

We shall now describe a basic training procedure for the reading of plans formulated on the basis of the preceding analyses.

This procedure is the result of a response[17] to a request for help from an organisation which undertakes vocational training for trades in the building industry, following acknowledgement of failure in this training. The goal set to us by the training organisation was to teach adults who had been seeking work for more than a year, and who were mostly of levels classified in France as 'alpha'[18] or 'post-alpha'[19], 'how to read plans for carrying out work in the construction industry.'

We set up the training process[20] around four types of situations:

• *communication* between consultants and contractors, in order to understand the functions served by technical drawings in human interactions;

-*copying* full-scale working drawings of basic figures in elementary geometry;

-*producing views* - the notion of view is introduced by a simple device which shows the orthographic projection on a reference plan. This device is made up of a pane of Plexiglas which can be written on, together with drawing implements (ruler, set square) and building-site tools (plumb line, spirit level). These implements allow vertical or horizontal transfers to be made from an actual object onto the Plexiglas[21]. This device had previously been the subject of experiments and checks designed for transfer activities, which were carried out in the meso-space[22]. Thus a view of an actual object *is not* what is experienced by seeing the object; it is a drawing similar to that which is obtained by transferring, vertically or horizontally, particular points linked[23] to the geometrical model of the object onto a reference plan (horizontal or vertical) which is represented by the pane of glass: the so- called 'indirect method'. Once introduced, the device remains available and is *a medium for validation*[24] (Margolinas, 1989) of reading and drawing which competes with and complements visual perception.

-*reading plans;* plans are read efficiently, but the technical problems of construction are not present: *interpretation of views is limited to the task of marking out.* Two simulations are made of the sites in which building workers operate. These allow graphics to be read in the meso-space[25] in the first one, and in the micro-space in the second.

It is these reading situations which will now be discussed.

5.1. Variables of basic plan reading

Our project consists in setting up certain conditions (variable values) necessary for the professional practices of reading[26]. The reading of plans in the training process is such that geometry acts as *a tool for effective marking out*, that is to say things are arranged so that certain *working conditions* on the construction site are reproduced. We shall describe a basic reading situation as follows:

The person carrying out the work is provided with a drawing. This drawing shows spatial relationships. The builder must mark out the contours on *actual flat surfaces*: he must decode the relationships and mark the outline upon the object.

The values given to certain variables in this reading situation correspond to different levels of difficulty in reading. During the process, the *size of the object* variable takes on two values, that of 'micro-space', or models of houses, and that of 'meso-space' which falls within the domain of the building-site, and which includes full size objects such as walls and ceilings.

> [...] the idea which students have of geometrical objects, the way in which they approach them, depends on their size [...]. The 'straight lines' and angles appear during the process of surveying in the *macro-space*. [...] The *micro-space* is the context in which small objects may be manipulated [...] The *meso-space* is the space within which the observer is able to see objects by moving around to gain different viewpoints. [...]. (Brousseau 1986, pp. 467-471)

Because of this and whatever the other features of the training situation may be, we maintain that the choice of an object size influences the way in which drawings are read as tools for marking out.

In this article we shall only describe reading situations in the micro-space (i.e. the space belonging to models) in order to illustrate the role played by a second variable, *position of the surface* (on which the marking out is to be done). During such a process, this variable takes on two values which are significant for reading, 'horizontal-vertical' (positions on the reference plan of orthographic projection) and 'oblique'. We have played with the values of this variable in the micro-space, in order to pass on to students the responsibility for formulating geometrical theories when working out an effective solution.

5.2. A teaching scenario for reading technical drawings within the micro-space of models

The students have three successive reading tasks in which the brief is always the same:

-each student receives a model of a house with garage (see figure 2) and three views of this model on a scale of 1 with the outline to be drawn (see figure 2);

-every student is asked to draw the outline on their model;

-the outline to be drawn is made up of a broken line which extends over two or three different surfaces depending on their position in the space.

So that the student may take responsibility for formulating the geometrical theories justifying the marking out technique, we organise things in such a way (by playing on the *position of the surface* variable) that the student has to make a *leap* in knowledge in order to solve the problem:

> [...] in order to make the student create a particular piece of knowledge, the teacher 'must' choose [variable] values which make this piece of knowledge optimal with respect to competing pieces of knowledge; progression is by leaps and not smooth. (Brousseau, 1997 p. 98)

We have organised the reading situations into a hierarchy seen from this perspective. We shall limit ourselves here to setting out the two opposed reading tasks, the intermediate task representing a small variation on the first.

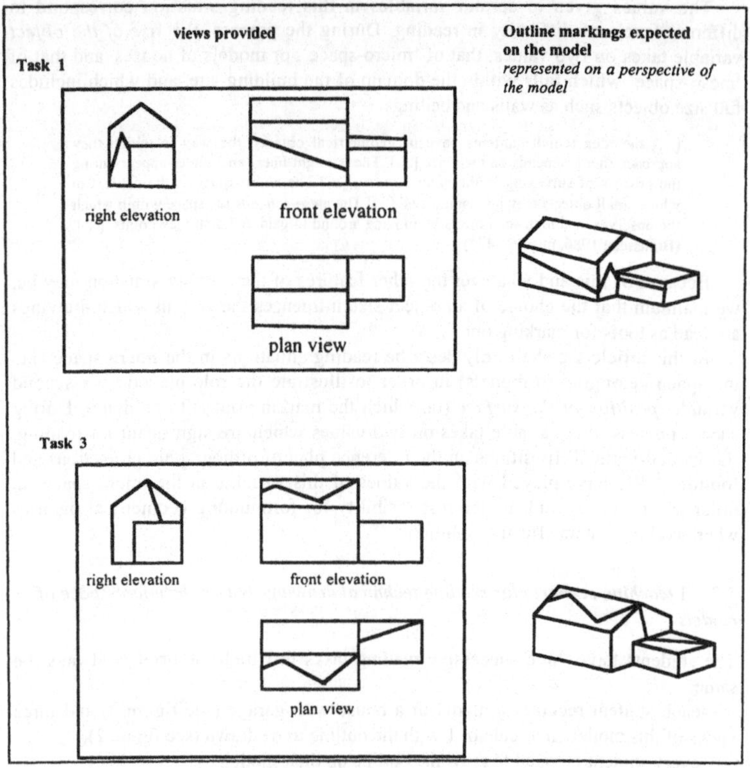

Figure 2. The graphics provided and the outline drawings expected

The consequence of choosing the micro-space is that the least 'costly' and the most reliable measurements are the measurements of straight lines and not the measurements of angles as would be the case in a macro-space. Appropriate techniques are thus those which are concerned with linking up measurements of lines, knowledge upon which the reading of technical drawings is built.

The presence of oblique plans in relation to projection plans (vertical or horizontal) is a feature of the object which allows these links to be differentiated.

What then are the appropriate drawing techniques for each of the two tasks?

In task 1, after having decided from the three views which ones relate to the outline to be drawn on which surface of the model, the techniques are the same as those for copying full-scale working drawings, that is to say in this instance:

-choosing relevant points on the drawing (ends of line segments for example);

-taking the necessary measurements (invariant in the relationship between drawing and surface);

-marking the corresponding points on the surface by using the measurements;

-joining the points.

A comparison of the measurements of line segments drawn on the model with those of line segments drawn on the views allows the validity of the drawings to be checked.

All the students manage to perfect a technique of this type.

In task 3, techniques involving either calculations of proportionality over the measurements or auxiliary constructions must replace the copying techniques. Only auxiliary constructions have appeared.

We have set out below extracts from the observation records of two students (called Bah and Men) who both begin by copying as in the preceding tasks. Faced with the failure of this first attempt, they look for an answer to the problem that has been set them in this task: Bah makes do with a pragmatic solution (appropriate in part) made possible by the Plexiglas, Men slowly develops a technique of a 'geometrical' type.

5.2.1. The case of Bah

Bah transfers the plan view onto the pane of glass. He places the pane of glass horizontally on the roof edge of the house-object and tries to transfer vertically onto the side of the roof an end of section of the plan view. The difficulty of doing this leads him to lay the pane of glass flat against the pitched pent of the roof and attempt to *bring together the rectangular surfaces of the house-object and the pane of glass*. He hesitates for a long time. He replaces his pane of glass horizontally and brings a first segment of the marked outline down onto the house-object. He finishes marking out by once more tipping the pane of glass over against the roof (a copying error).

5.2.2. *The case of Men*

Where Men sees the task as problematical

Men transfers the drawing of the front view onto the Plexiglas. He reproduces this drawing by copying it just as it is on to the pitched pent of the roof. Next he erases it. He then draws on the model the outlines of the horizontal and vertical plans. The intervention of the teacher (called T) leads him to explain the problem.

T: Well then Men?

Men: That *(the roof)* - that's a problem. I've done this (*he shows the front view on the pane of glass).*

Men indicates that he wants to take the height from the edge of the roof on to the front view and says:

Men: there's a question of perspective

T: Why?

Men: We're sure that the length's right.

T: And the height?

Men: There's a problem with the height

T: How can you be sure of the height? How can I transfer a height and be sure about it? How do you do it in concrete terms?

Men The plan view that's distorted as well. There's no view which gives us the height.

T: Can't you find the height by using a flat surface? (*He shows the left facade of the house)*

Where Men starts to develop a technique for marking out

Men adjusts the pane of glass against the left facade of the house. He draws on the left facade a vertical section which is the same length as the height given on the front view: he obtains point A1, the end of the outline on the left pitched edge of the roof.

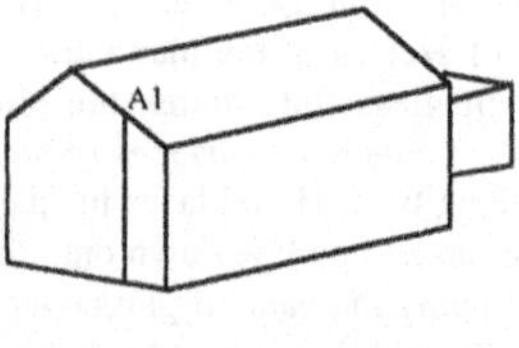

He draws on the pent of the roof two parallels to this edge by using 'horizontal' measurements (taken from the plan view) corresponding to the ends of the marked out segments.

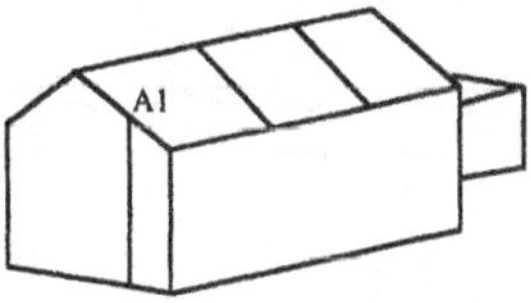

Where Men falls back on a copying technique

Men takes the measurement of the first segment from the plan view: making adjustments with the help of his graduated ruler, he decides on a point A2 on the first parallel, whose distance from A1 is this measurement. By repeating this process, Men obtains point A3 which he joins to the right end of the edge of the roof.

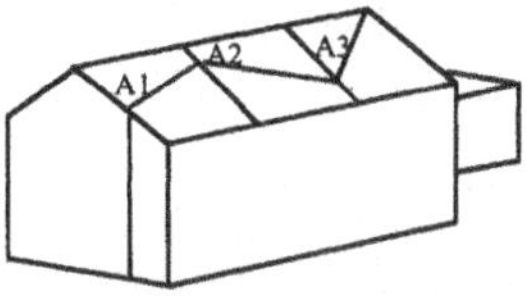

Where the teacher judges by checking the marking out against the pane of glass

The teacher checks the result against the pane of glass and judges:

T: There is a difference, even so.

Men explains his method.

T: Are you sure?

Men: Yes, because the lengths aren't dwarfed

T: Horizontally, OK. But diagonally it doesn't work. Can't you do for A2 what you've done for A1?

Men: Ah! Yes.

Where Men concludes development of a marking out technique

Men ends by marking out lines showing the level on the left-hand side, for which he has measured the height on the front elevation. He extends these lines showing level onto the pent of the roof: he obtains the ends of the sections by intersecting them with the parallels to the pitched edges.

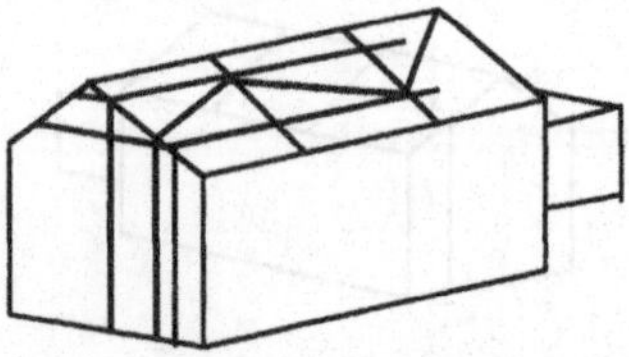

Men's exemplary success nevertheless required the intervention of the teacher for him to take into account the effects of linking up the measurements. For the construction of point A1, the invariants (vertical and horizontal) of measurement can function separately. For the construction of points A2 and A3, an extra difficulty is added, that of co-ordinating the horizontal and vertical invariants within the linking up of measurements - which explains the reappearance of the copying error. The *auxiliary outline markings* of Men's technique show these invariants in terms of graphics. It was this technique which would then become institutionalised, Men taking on the role of teacher for a while.

The cases of Men and Bah illustrate the contribution of the pane-of-glass-tool to the situation: it offers the trainer the possibility of not judging directly what the students have produced but enabling the students themselves to make a personal appraisal of the validity of their marking out (the glass becomes a medium of validation). In the particular case of Bah, this device allowed him to avoid failure by providing him with the elements of a pragmatic solution.

6. CONCLUSION

In this article, we have given prominence to two types of links with geometrical knowledge:

-in the training process, geometrical knowledge can be part of a 'discourse' which allows the teacher (who is not a mathematician) to justify the practices taught other than by arguments based on authority or on practical efficiency: such a 'discourse' enables the teacher both to respond to a student's desire to understand, and to take on the role of teacher[27], more knowledgeable in subject than the students. But in basic vocational training in France, there is a tendency to eliminate it from exercises, which are the student's primary focus of work.

-in practice, geometrical knowledge is working knowledge for developing solutions to problematical tasks - marking out, for example.

Our procedure attempts to consider technical drawings as a geometrical model (in relation with orthographic projection) whilst at the same time maintaining relevance to the building-site situation. In this process, the student implicitly (via some pragmatic solution) or explicitly (via auxiliary outline markings) uses geometrical knowledge (invariant of orthographic projection) for part of the work, in order to develop techniques for marking out.

One question remains: that of the likelihood of adopting such a process, which goes against present educational trends in France. One needs to ask: why are these trends emerging? What particular limitations of vocational training in different institutions do they stem from? For us, these questions belong to the field of educational research: they remain open.

7. APPENDIX: ORGANIZATIONAL DIAGRAM OF THE FRENCH EDUCATION SYSTEM

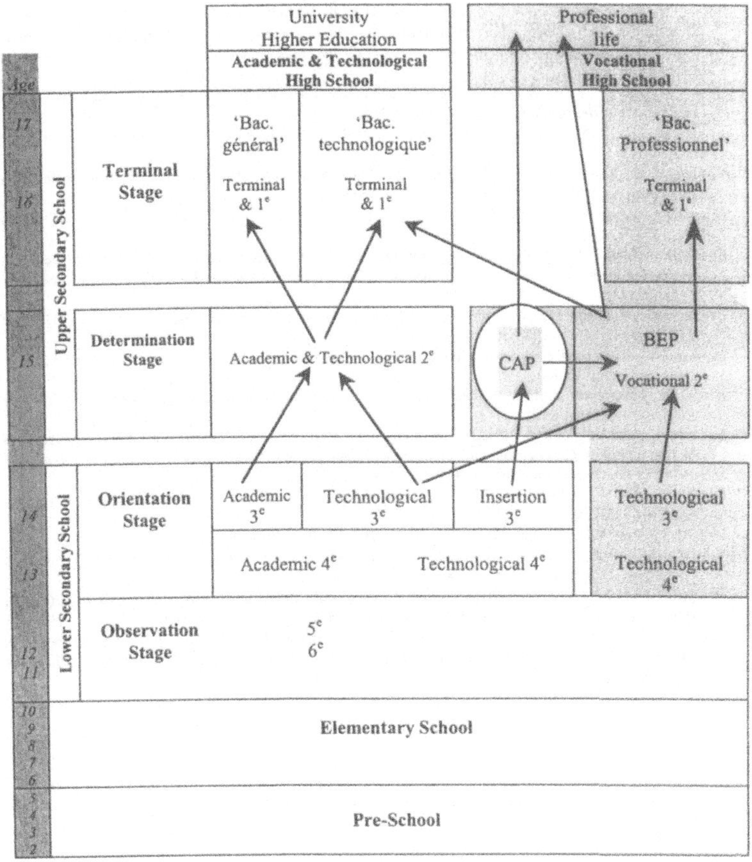

8. REFERENCES

Arsac, G. (1987). L'origine de la Démonstration: Essai d'Épistémologie Didactique. *Recherche en Didactique des Mathématiques, 8*(3), 267-312.

Artigue, M. (1990). Ingénierie Didactique. *Recherche en Didactique des Mathématiques, 9*(3), 281-307.

Berthelot, R., and Salin, M.-H. (1992). *L'enseignement de l'Espace et de la Géométrie dans l'Enseignement Obligatoire.* University of Bordeaux I.

Bessot, A., Déprez, S., Eberhard, M., and Gomas, B. (1992). *Approche Didactique des Processus de Formation de Base à la Lecture de Systèmes de Vues, Destinés à des Adultes Peu Qualifiés dans le Cadre des Métiers du Bâtiment.* Grenoble: IMAG.

Bessot, A., Déprez, S., Eberhard, M., and Gomas, B. (1992). Le Sens des Graphismes Techniques dans la Formation Professionnelle de Base pour les Travailleurs du Bâtiment: Une Approche Didactique. *Education Permanents, 111*, 87-105.

Bessot, A., Deprez, S., Eberhard, M., and Gomas, B. (1993). Une Approche Didactique de la Lecture des Graphismes Techniques en Formation Professionnelle de Base aux Metiers du Batiment. In A. Bessot and P. Verillon (Eds.), *Espaces Graphiques et Graphismes d'Espaces* (pp. 115-144). Grenoble: La Pensee Sauvage.

Bessot, A., and Eberhard, M. (1983). Une Approche Didactique des Problèmes de la Mesure. *Recherches en Didactique des Mathématiques, 4*(3), 293-324.

Brousseau, G. (1986). *Théorisation des Phénomènes d'Enseignement des Mathématiques.* University of Bordeaux I.

Brousseau, G. (1997). *Theory of Didactical Situations in Mathematics* (Vol. Edited and translated by Balacheff, N., Cooper, M., Sutherland, R., and Warfield, V.).

Chevallard, Y., and Julllien, M. (1991). Autour de l'Enseignement de la Géométrie au Collège. *Revue 'Petit X', 27*, 41-76.

Chevallard, Y. (1994). Integration et Viabilite des Objets Informatiques dans l'Enseignement des Mathematiques. Le Probleme de l'Ingenierie Didactique. In Cornu (Ed.), *Nouvelle Encyclopedie Diderot.* PUF.

Chevallard, Y. (1996). *Séminaire d'Analyse et d'Ingénierie Didactique* (internal document): IUFM of Aix-Marseille.

Deforge, Y. (1981). *Le Graphisme Technique, son Histoire et son Enseignement.* Collection Milieux: Editions Champ Vallon.

Klein, F. (1896 (new edition Vuibert, Paris 1931)). *Leçons sur Certaines Questions de Géométrie Élémentaire.* Nouy, Paris.

Laborde, C., and Capponi, B. (1994). Cabri-géomètre Constituant d'un Milieu pour l'Apprentissage de la Notion de Figure Géométrique. *Recherches en Didactique des Mathématiques, 14*(1-2), 165-210.

Monge, G. (1796 (new edition Jacques Gabay, Paris 1989)). *Géométrie Descriptive.*

Olivier, E. (1976). *Implantations, Tracés, Nivellement, Relevés : Travaux Pratiques.* Paris: Entreprise Moderne d'Edition.

Rabardel, P., and Weill-Fassina, A. (1987). *Le Dessin Technique.* Paris: Hermès.

Vérillon, P. (1996). La Problématique de l'Instrument: un Cadre pour Penser l'Enseignement du Graphisme. *Graph and Tec, 0*, 57-78.

Vergnaud, G. (1996). The Theory of Conceptual Fields. In L. P. Steffe, P. Nesher, P. Cobb, G.A. Goldin, B. and Greer, B. (Eds.), *Theories of Mathematical Learning.* Mahwah: Lawrence Erlbaum.

Weill-Fassina, A., and Rachedi, Y. (1993). Mise en Relation d'un Espace Réel et de sa Figuration sur un Plan par des Adultes de bas Niveau de Formation. In A. Bessot and P. Verillon (Eds.), *Espaces Graphiques et Graphismes d'Espaces.* Grenoble: La Pensée Sauvage.

9. NOTES

[1] this article is based on research carried out by the following team: Bessot, Déprez, Eberhard, & Gomas

[2] through Euclidean geometry.

[3] an object in space may, for these purposes, be a building.

[4] Rabardel & Vérillon (1993), Vérillon (1996) speak of the 'instrumental' aspect of technical graphics.

[5] Depending on the scale of the project, the person 'setting out' could be the foreman, a surveyor, the site engineer, or any one of a number of workers in the construction industry. In the text, we refer to this person simply as 'the worker'.

[6] 'This consists of showing in succession, on the piece of land earmarked for the work: a) the boundaries of the property; b) the contours of the general excavation where building is to take place; c) the lines and contours of the actual digs and of the foundations, in other words of those elements which will bear the weight of the building.' (Olivier 1976, p. 18-19)

[7] 'This consists of showing, by means of actual lines, the contours of the work or part of the work to be carried out, with their true dimensions. This shall no longer be done on the soil, but: - either on flat areas of paving or floors [...] - or on vertical or non-vertical partitions - or on building materials waiting to be cut out.' (Olivier 1976, p. 19)

[8] This is a reference work designed for practitioners on the building-site, future technicians in these professions and those responsible for vocational training for the building-site.

[9] We only describe some of these procedures, in 'geometrical' terms.

[10] or multiples of these wholes according to the size of the outline drawing to be made.

[11] an example may be found of this in the article by Bessot in section 4 of this book.

[12] Deforges traces this development from Monge's descriptive geometry (1796) to the present day.

[13] that is to say definable in terms of objects of Euclidian geometry.

[14] see Laborde & Capponi (1994) for another diagram relating to *geometrical figures.*

[15] 'drawing' here means all the procedures involved in carrying out a drawing.

[16] By the term 'didactical engineering' we mean a form of didactical work 'comparable to the work of the engineer who, in order to realise a specific project, relies on the scientific knowledge of his field, and agrees to submit to checks of a scientific nature, but, at the same time, has to work on subjects which are much more complex than the refined subjects of science and therefore has to tackle on a practical level, with all the means at his disposal, problems which science cannot or will not yet take on.' (Artigue, 1990)

[17] the Bessot, Déprez, Eberhard & Gomas team assumed responsibility for this.

[18] illiteracy in French and difficulties with spoken French (people often not of French mother tongue); inability to master basic calculations (addition and multiplication).

[19] difficulties with reading and writing French, particularly with understanding texts; mastery of addition and multiplication but difficulties with other operations.

[20] The process produced by our didactical engineering was tested four times: twice during adult training and twice in a technological class for 13-14 year olds (see Appendix). The courses for adults were refresher courses aiming to restore vocational skills to a level which would enable them to find work. For a detailed description of the engineering, see Bessot et al 1992.

[21] the pane of glass can be handled and arranged with ease.

[22] see below in this article.

[23] link founded by the theorem-in-action (Vergnaud 1996, see also in this volume): 'in order to draw a line segment it is necessary to build at least two points belonging to this segment'

[24] '[...] in order that learning may continue during those stages when the teacher himself ceases to supply information directly, it is necessary for the medium to allow stages of validation.' (Margolinas, 1989) Such a medium is a medium for validation.

[25] See below in this article.

[26] This is different than Hanh's project (see this section), which is to create situations using one's own professional practices, i.e. 'personalized situations'.

[27] who since Monge has used geometrical knowledge to justify outline drawings without however linking it to descriptive geometry.

CHAPTER 11

TEACHING MATHEMATICS TO SHOP-ASSISTANT APPRENTICES EXPLORING CONTENT AND DIDACTICAL SITUATIONS

CORINNE HAHN

Abstract. This article presents the results of a research study conducted with apprentices in order to examine the double hypothesis that vocational training incorporating work placement gives the opportunity to construct authentic situations out of the everyday practices of students and that these situations allow students to improve their understanding of mathematical problems. After analysis of workplace situations had shown that the only mathematical concept used by apprentices was that of proportionality applied to percentage calculations, we first analyzed the way in which students at several levels handled this idea of percentage. We then devised an experimental framework in order to help the apprentices to transfer their knowledge.

In the past years, many research studies explored out-of-school mathematical practices (see Adda, 1975, Cerquetti, 1981, Lave, 1988, Nunes et al, 1993). These studies showed the existence of informal mathematical procedures, most of the time without errors, sometimes complex but, in all cases, far removed from school procedures which seem difficult to transfer.

The results of the research studies lead us to wonder about the methods that could help students to make sense of mathematical content at school and make them capable of applying their knowledge out of school (Boaler, 1993, 1994). But is it really possible to bring the 'real world' into the mathematics classroom (Sierpinska, 1995)? And, if it is possible, will it really facilitate transfer?

This article presents the results of a research study conducted from 1993 to 1995 with jewel shop-assistant apprentices[1] in order to answer these questions.

1. MATHEMATICS USED BY A JEWELRY SHOP ASSISTANT

The jewelry profession uses a very limited set of mathematical notions: In fact, proportionality is almost the only concept the apprentices must apply, in particular, for calculations of gold (Karats) and precious stones (carats). The most frequent application of proportionality is the percentage within the framework of 'price

Bessot & Ridgway (eds.), Education for Mathematics in the Workplace, 159—166.

calculations'. It implies the use of a very broad context which is quite embarrassing: Up to what point should we be teaching this context in the mathematics classroom?

Difficulties related to percentage calculations not sufficiently accounted for (Hahn, 1995) are amplified here by the differences in the discourses: Words can have different meanings, depending on the situation in which they are used.

A case in point is the word *coefficient*.

-In the mathematics classroom, it has the general meaning of a proportionality operator between two quantities linked by a percentage operation;

-In the management classroom a few 'coefficients' are identified by 'formulas';

-For shop owners, the 'coefficients' is the multiplicative operator between their purchase price (exclusive of tax) and their sales price (inclusive of tax).

Procedures are also very different: At school, fractional operators are preferred, while in the shop, symbolic procedures ('- x%' or, most of the time, the inaccurate procedure '± x%') are mostly used.

For example, to calculate a *pre-tax*[2] at school, the apprentices are taught to perform × 100 ÷ 118.6 or ÷ 1.186 while shop owners prefer to perform '- 15.68%', × 0.84327 or ÷ 118.6%[3].

2. THE MATHEMATICS COURSE

The BEP[4] curriculum contains a heavy mathematical content. The content tends to be an approximation of that which is taught in high school. Necessarily, it has very little in common with reality in the work place, because of the restricted corpus of mathematical knowledge used in the shop.

In the case of the jewelry shop-assistant apprentices, the mathematics classroom is not very different from the usual BEP mathematics classroom, except that it focuses on word problems the context of which is related to the professional background of the students.

If the apprentices prefer to solve problems relative to jewelry, it does not really help them to understand mathematics. Most of the time, mathematics teachers are unaware of their professional reality, there are very few contacts between them and the shop owners, even with the technical teachers: Teachers only present 'classical' problems embedded in a pseudo-professional context, and students who believe in their 'reality' are usually disappointed.

Here is an example of what can happen. A teacher had asked a student to calculate the discount given to a customer according to this scale:

Quantity Purchased (FF)	Discount Percentage
0 to 2,000	0
2,000 to 4,000	2
4,000 to 8,000	3

A shop owner whose apprentice had a bad grade because she had calculated 3% of FF5,000, explained to us that

> the teacher's solution was wrong because it never happened like that in the shop: If a customer buys for an amount of FF5,000, I must give him a 3% discount. If I do as the teacher says[5], the customer can sue me.

It was clear that, by proposing a situation falsely professional, rejected by the shopowner, the mathematics teacher had lost any credibility he could have with the shop owner and his apprentice (on the contrary, the 'guelte', the commission given to shop assistants in proportion to their sales volume, could effectively be used for such a calculation).

In this case, the teacher could have created a more realistic problem, if he or she had had a better knowledge of the professional context. However, it is not always possible since word problems usually present expurgated solutions, with very little background 'noise'. It is very difficult to offer real hard 'concrete'.

3. STUDENTS' SKILLS WITH PERCENTAGES

In order to analyze the way in which students at several levels handled this idea of percentage before designing an experimental framework, we conducted a study of three student groups, as follows:

-The first group of 111 apprentices solved seven word problems about percentages. Some of them answered twice, first in the mathematics classroom, and then six months later in the technology classroom;

-The second group of 162 students at the beginning of the BEP curriculum answered a multiple-choice questionnaire;

-The third group of 221 students at the end of the BEP curriculum were asked to answer the same test. Some of them responded twice, the second time (during which they had to explain their responses) three months after the first one.

This study brought several facts to light.

We first noticed that there were two different practices, depending on the environment in which the apprentice needed to solve the problem. For example, the same student used a different procedure to solve the same problem depending on whether he or she was in the mathematics classroom or in the technology classroom. In the technology classroom, the student instinctively uses the procedure of the shop. The student considered the school practice only as a second choice when in absence of a professional practice.

If percentages seem more familiar to students at the end of the BEP curriculum, it does not mean that they have actually understood the notion. Very often, students have a very confused idea about what a percentage is.

An apprentice, who had to calculate the new price with a 25% discount, had written:

$$800 \times 75 = 600$$

I asked her to reconsider her response. Then she wrote:

$$800 \times 75 = 600F$$

Feeling that I was very surprised, she shrugged her shoulders and wrote:

$$800 \times 75\% = 600F$$

She obviously only added the symbol to please her teacher.

I also realized that apprentices did not succeed any better than full-time BEP students. They often mix school practices with shop practices, as follows:

$$\frac{2700-20}{100} = 2699.80 \qquad \frac{2700 \times 20\%}{100} = 540$$

In both cases, it seems that apprentices brought their experience from the shop into the classroom. I asked the first student to explain to me how he had proceeded. He took his calculator and entered *2700 - 20%*. When he saw the result, he told us that when he had solved the problems, he had forgotten his own calculator and had borrowed his neighbour's. 'The calculator was wrong', he explained...

Apprentices often use procedures that they do not understand.

A shop assistant wrote that, to determine the price with a 20% discount of an object normally sold at FF250, she performs $250 \times 8 = 200$. She explained to me later that the owner of the shop where she worked told her to multiply by 8 and then read the three first numbers on her calculator. This discount was given for only a certain type of watch, the price of which was always below FF1,000. The shop assistant later appeared incapable of solving the same problem for a price of FF2,700.

I also noticed a few calculation reflexes among the three student groups, in particular:

-The well-known 'additive percentage procedure'. Students think that '+ x%' is the inverse operation of '- x%' (see Freudenthal, 1983). They apply a mathematical procedure (the inverse operation of '-' is '+') to a familiar expression that they believe is mathematical, forgetting that a percentage is always associated with a subset of reference. For example, to calculate the price without tax (tax 18.6%), they enter on their calculator 'price tax included - 18.6%', and they do not take into account that the amount on which the percentage of tax is calculated is the price without tax and not the price tax included.

Only 21% of the apprentices knew how to calculate the price without taxes from the price with taxes included (tax at 18.6%), while 44% of them used the wrong procedure, that is '- 18.6%'. The results were almost the same when they had to calculate the price before the discount, while knowing the price which includes a

30% discount. Results were 29% correct responses and 46% incorrect, due to the application of the incorrect procedure '+ 30%'.

Only 24% of the students at the beginning of the BEP curriculum and 36% of the students at its end thought that if an article is sold at FF400 with a 10% discount, to calculate the original price, they needed to divide FF400 by 0.9.[6]

-The more surprising 'net price procedure'. Students no longer know how to calculate the discount itself (or the tax), but systematically calculate the new price, with the discount or the tax included.

Ninety-nine percent of the apprentices were able to calculate the new price after a discount, and 94% to calculate the price including taxes. However, only 58% of the same apprentices answered correctly when they were asked to calculate the amount of discount, and 64% for the amount of tax.

Also, one fifth of the students entering BEP or the Bac Professionnel thought that the statement '20% = 0.8' was true.

But apprentices seem more aware of what they are doing when the school practice and the professional practice resemble each other. For example, the procedures are very stable in the case of the calculation of the price with a discount, where the decimal operator is preferred at both school and in the shop.

4. SUITABLE DIDACTICAL SITUATIONS

What type of situation will help apprentices transfer their knowledge from one educational environment to another?

Apparently, the use of 'contextualized problems' does not help students, but we also observed that it was not hard concrete.

We first worked on a 'real situation' as opposed to a 'realistic situation'. Called the 'Gonsard Case', it was built on a genuine professional situation and used the vocabulary and procedures observed in the shops. It consists of a story and a collection of genuine documents, including an invoice, a poster of the professional federation, and clippings from professional newspapers.

Students first checked the authenticity of the situation, then became heavily involved in the activity. But however real the situation, it was still the teacher's situation and not the students' (Geay, 1994).

So we decided to construct a framework within which the students could prepare their own problem. This was done with an interdisciplinary project related to their professional practice: Each apprentice was asked to collect information about his shop, its environment and a jewel chosen by his or her shop owner.

The mathematics course on percentages was built from this information, from these 'personalized situations', reversing the usual progression. The traditional course started with revisions on rational numbers, then went on to the calculation of percentages, and, finally presented the application of 'price calculations'. We, however, worked the other way around, starting with a debate based on the information the apprentices had collected, to determine a common meaning to the terminology they used. The class then worked on all the personalized problems, such as:

A group helped Séverine to calculate the net price of her jewel. Three students went to the board to explain their solutions.

Student 1: *net price = 1407.76   1497.62 - 6%*

Student 2: $\frac{1497.62 \times 6}{100} = 89.8572 = 89.86$

1497.62 - 89.86 = 1407.76

Student 3: *100 - 6 = 94* which becomes *0.94*

1497.62 × 0.94 = 1407.77F

Student 4: *becomes ?*

Student 3: *94 divided by 100...*

Teacher: Why?

Student 3: *???*

Student 1: The second method is simpler.

The group soon understood that being able to use a variety of methods is extremely valuable. One method gave the net price while another one gave the amount of discount as well as the net price. The teacher was also able to explain why the procedure written by Student 1 was wrong.

To attempt to determine the influence of the latest experiment on the comprehension of the notion of the percentage, we carried out a dual evaluation with apprentices who participated in the project, using as indicators the two false procedures described above.

-We first measured the development of their skills during the school year, and made a comparison with the development observed for other apprentices;

-we also compared the results of this group of apprentices with the results observed in the past with other jewelry shop-assistant apprentices.

For apprentices involved with this experiment, we observed a significant decrease of 'net price' reflex and the stabilizing of procedures. This seems to indicate a better identification of mathematical structures, behind the context. For example, students use the same procedure to solve a tax-related problem as a discount-related problem. Apparently, apprentices found sense in their practices.

But if procedures were more stable, they were not for that reason more frequently correct: We observed no significant improvement concerning 'additive percentage' procedures. Problemitizing situations from their own experience did not seem to have helped apprentices to assimilate the underlying mathematical concept.

5. CONCLUSION

Introducing reality into the mathematics classroom does seem to help apprentices to make sense of the notion of proportionality applied to percentage calculations, but it does not help them go beyond what we believe to be the main obstacle : 'additive percentage ' procedures.

Still, we must take into account the fact that this experiment was only a three-month parenthesis in a curriculum that was still traditional, which means that it was a curriculum designed to meet requirements of the final exam.

Anyway, in the event that the experiment had been a complete success, it would have been impossible to transform the entire curriculum into 'personalized situations', because the mathematical content and the professional reality have very little in common. Should we eliminate the greater part of the mathematical content from the curriculum? Surely not: A professional education cannot be restricted to a limited number of basic skills if we do not want to mortgage the future of our students. Mathematics cannot be reduced to its utilitarian aspect alone.

We believe that the use of a 'personalized situation' at the start of the curriculum, in a interdisciplinary framework, will help modify the interactions inside the mathematics classroom, and open it to the apprentices' own 'ethnomathematics' (see Boaler, 1994).

6. REFERENCES

Adda, J. (1975). L'incompréhension en Mathématiques et les Malentendus. *Educational Studies in Mathematics, 6*, 311-316.

Boaler, J. (1993). Encouraging the Transfer of School Mathematics to the Real World through Integration of Process and Content, Context and Culture. *Educational Studies in Mathematics, 25*, 341-373.

Boaler, J. (1994). The Role of Contexts in the Mathematics Classroom: Do They Make Mathematics More "Real"? *For the Learning of Mathematics, 13*(2), 12-17.

Cerquetti, F. (1981). *Quelques Aspects de la Relation aux Mathématiques chez des Élèves de LEP et Classes Pratiques.*, Université Paris 7, Paris.

Freudenthal, H. (1983). *Didactical Phenomelogy of Mathematical Structure*. Dordrecht: Reidel.

Freudenthal, H. (1991). Revisiting Mathematics Education. *Mathematics Education Library, 9*.

Geay, A. (1994). *Le Systeme Alternance*. Paris: l'Harmattan.

Hahn, C. (1995). *Le Lien Mathématiques/Réalité dans un Enseignement en Alternance. Le Cas du Pourcentage dans une Formation à la Vente.*, Université Paris 7.

Nunes, T., Schliemann, A., and Carraher, D. (1993). *Street Mathematics and School Mathematics*. Cambridge: Cambridge University Press.

Sierpinska, A. (1995). Mathematics: "In Context", "Pure", or "With Applications"? A Contribution to the Question of Transfer in the Learning of Mathematics. *For The Learning of Mathematics, 15*(1), 2-15.

7. NOTES

[1] These apprentices alternate between one week in a jewelry shop and one week at school. They prepared the *Brevet d'Etudes Professionnelles*, BEP, a French national, two-year professional curriculum prepared in either a CFA (cooperative education) or in a professional high school (traditional school system).

[2] Find the price not including tax, while knowing the price with tax included and the tax rate (18.6%).

[3] To find the price with a 18.6% tax included, one has to multiply the price not including tax (x) by 1.186 (x + 0.86x). Then, in order to find x, knowing the price with tax included, he or she must divide by 1.186 which is approximately the same than multiply by 0.84327 (=1÷1.186) or apply a reduction of 15.68% (1-0.84327 ≈ 0.1568).

[4] see note 1

[5] For example, for a purchase of FF5,000 the discount would be:

0% of FF2,000	FF 0	
2% of FF2,000	FF 40	
3% of FF 1,000	FF 30	Total: FF 70

[6] It was one of the questions of the test in which they had to answer 'right', 'wrong', or 'I don't know'.

CHAPTER 12

DEVELOPING A NEW MATHEMATICS CURRICULUM FOR POST-COMPULSORY EDUCATION

GEOFF WAKE, JULIAN WILLIAMS

Abstract. We describe the development of a curriculum construct 'general mathematical competence' which we have used in curriculum development of teaching materials in vocationally focused post-16 courses. This construct is designed to bridge the academic/vocational divide in the U.K. education system. It is now being used to define an *assessment and accreditation* system for applied mathematics units which are intended to attract students from both sides of this divide to study mathematics as an option.

1. INTRODUCTION

Our future workforce will increasingly need to be able to cope with complex technical problems and rapidly changing working practices. Mathematical competence will therefore be a foundation on which training will need to build. A new construct, general mathematical competence, that allows organisation of mathematics in pre-vocational courses is described here. The idea of general mathematical competence has been developed further in specifying new mathematics qualifications that attempt to bridge the academic-vocational divide. These aim to encourage more students to apply mathematics to their current studies or future work.

This approach had its genesis in pre-vocational courses, where the need for mathematical competence is often obvious. However the most appropriate means of supporting students recruited to these courses is not always clear. The momentum that has taken the development forward to the point of specifying new qualifications has come from the recognition that the UK is unusual in not demanding some form of mathematical study of post-16 students and the perceived need for appropriate optional qualifications that will prove attractive to these students.

2. PRE-VOCATIONAL COURSES

Pre-vocational courses that are primarily aimed at the 16-19 age cohort are being organised in England and Wales into a common framework: General National

Bessot & Ridgway (eds.), Education for Mathematics in the Workplace, 167—180.

Vocational Qualifications (GNVQs). These have been modelled on 'true' vocational qualifications (NVQs) which credit competence in vocational practice. The GNVQ on the other hand is used in schools and colleges primarily for full-time students who wish to progress either directly, or via Higher Education, to work in a particular vocational field such as 'Health and Social Care', 'Leisure and Tourism' or 'Engineering'. For the first time in England and Wales mathematics has been identified as a core part of all these pre-vocational qualifications: specifications exist for expected competence in the Key Skill of 'Application of Number', across every GNVQ, at every level, alongside those for the other Key Skills of Communication and Information Technology.

The GNVQ serves to provide a general education with a particular vocational bias for the target cohort. This 'general' aspect of pre-vocational courses is recognised as having dual value for progression either into Higher Education or directly into employment. The possibility of this duality of GNVQs becoming part of other qualifications at advanced level and hence facilitating transfer between 'academic' and 'vocational' strands was recognised by Jessup (1991). However each of the 'user' groups of the qualification, Higher Education and Employment, has a different emphasis in their support for mathematical competence. The education community requires attention to the internal coherence of subject content and development of mathematical understanding. Additionally from this group there has been an increasing demand for students to develop mathematical problem solving and modelling skills. On the other hand employers often call for students to be able to use mathematics to solve problems in vocational contexts; from this group the emphasis is often on technical fluency and competence.

Naturally there are different emphases in terms of quantity and focus of the mathematical demands made of students across the range of vocational areas (see for example Sutherland and Pozzi, 1995). Many students do aspire to progress to study in Higher Education as the work of Wolf (1994, 1997) ascertained and much effort has been made into researching the mathematics demands of Higher Education, (see for example Lord, Wake and Williams (1995)).

Those students working on 'non-technical' courses will generally find that competence demonstrated in *'Application of Number'* will meet the demands of Higher Education. However, students following courses in the technology cluster are expected to undertake study of mathematics well beyond this and indeed at a level comparable with that in 'academic' A Level Mathematics courses.

Teachers currently involved with the delivery of pre-vocational qualifications in schools and colleges support our contention that their students (even many on technology based courses) have little or no intrinsic motivation to study mathematics itself, but rather have a potential willingness to invest time and energy on acquiring mathematical competence if it can be seen to enhance their technical capability in their chosen vocational field. The willingness of students to engage with mathematics in this way is evidenced by the accounts of the different trainees described in the other papers of this section.

A further problem that we acknowledge is that of the ability of students to 'transfer' their mathematical knowledge, skills and understanding, both in terms of transfer from the mathematics classroom to the workplace and even between the mathematics classroom and other classroom situations (see Ernest (1998) for a background to how current theories attempt to explain this problem). We attempt to take account of this thorny problem by encouraging students to see the value of using mathematics in their chosen 'vocational' field while also recognising its generality; hence the point of the term 'general mathematical competence'.

3. GENERAL MATHEMATICAL COMPETENCE.

In an attempt to resolve the conflicting demands of a general mathematical education for progression and mathematical competence in vocational situations, while recognising that the potential motivation for students lies in the latter, our initial strategy was to develop the new construct 'the general mathematical competence'. Each *general mathematical competence* organises a corpus of relevant mathematics (models, concepts and skills) and demonstrates how this can be applied to successfully solve problems in a range of relevant contexts. Each concentrates on mathematical models as being useful to understand the real world, and highlights the limitations of the models used as well as their generality.

Our initial work was based in the area of Science at Advanced Level. This was supported by the curriculum development project, Nuffield Science in Practice, and the Department for Education and Employment (DfEE). The *general mathematical competences* developed in response to asking the question 'What mathematics do science students use and how do they use it?' included:

-Handling Experimental Data Graphically
-Using Models of Direct and Inverse Proportion
-Interpreting Large Data Sets
-Using Formulae
-Measuring in Science
-Using Mathematical Diagrams
-Costing a Project.

These were developed by considering the need for students to not only use and apply mathematics as science students but also to demonstrate competence across the full *Application of Number* specifications. We believe other students following pre-vocational qualifications will find that these *general mathematical competences* are applicable in their particular area of study. This may be relatively easy to demonstrate within the technology cluster of qualifications but requires further research into other vocational areas.

Modular materials have been developed to support delivery of mathematics using *general mathematical competences*. Each Module

-has an introduction using case studies, illustrations and/or practicals from the GNVQ Science course which can clearly be seen to be generalisable to other relevant applications (for further details see Wake & Williams, 1995);

-teaches an identifiable body of mathematical knowledge, (facts, concepts and skills) from the *Application of Number* syllabus and where necessary beyond;

-explicitly teaches the mathematics as a set of adaptable models, wherein the specification of assumptions, interpretation and validation are critical;

-identifies potential points where remedial help may be needed, and cross references support materials;

-identifies starting points for interesting mathematical inquiries which may take the student beyond the minimal requirements of 'competence'.

Underlying mathematical concepts are used in more than one general mathematical competence. For example, proportional reasoning is used in the general mathematical competence 'Using Models of Direct & Inverse Proportion' in both graphical and algebraic forms to develop models that explain scientific phenomena, while also underpinning much of the work involving scale drawing and representation in the general mathematical competence 'Using Mathematical Diagrams'. It is hoped that students will develop an understanding of how a basic concept, in this example proportional reasoning, underpins a number of applications of mathematics.

4. EXEMPLIFYING THE GENERAL MATHEMATICAL COMPETENCE IN STUDENTS' WORK IN SCIENCE

A general mathematical competence that clearly fits with the study of 'vocational' science is that of 'Handling Experimental Data Graphically'. In their study of the vocational science units one would expect students on many occasions to use graphs to help them make sense of observed scientific phenomena. Figure 1 below, shows the output of a student working in this way. Here, the student investigates graphically whether or not the resistance of a wire is directly proportional to its length.

The data is competently plotted with some attention being given as to whether or not a straight line can be fitted through it. The line identified by the student as being the 'best straight line' indeed passes through the origin and demonstrates that the direct proportion model under investigation has some validity. However, the model proposed by the student as the 'worst straight line' is not investigated for proportionality. The student's work suggests confusion between the idea of proportional and linear models. Careful inspection shows that the student has been inaccurate in her calculations. When calculating the gradient of the 'best straight line' she uses point A, which does not lie on this line; when calculating the gradient of the 'worst straight line' she indicates that she is to use co-ordinates of points on the 'best straight line and even uses incorrect values for these.

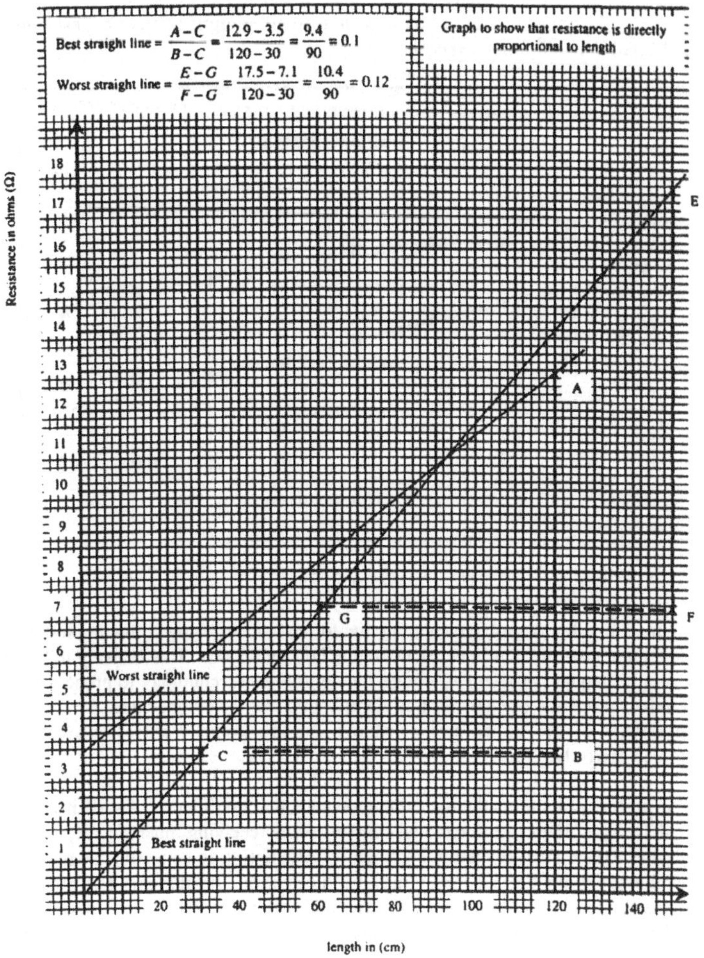

Best straight line = A−C / B−C = 12.9−3.5 / 120−30 = 9.4 / 90 = 0.1
Worst straight line = E−G / F−G = 17.5−7.1 / 120−30 = 10.4 / 90 = 0.12
Graph to show that resistance is directly proportional to length
Resistance in ohms (Ω)
length in (cm)
Worst straight line
Best straight line
A
B
C
E
F
G

Figure 1.

A further example of a student's work with linear models is shown in Figure 2, below. This illustrates an attempt by a student to fit straight lines to experimental data collected in two titration experiments.

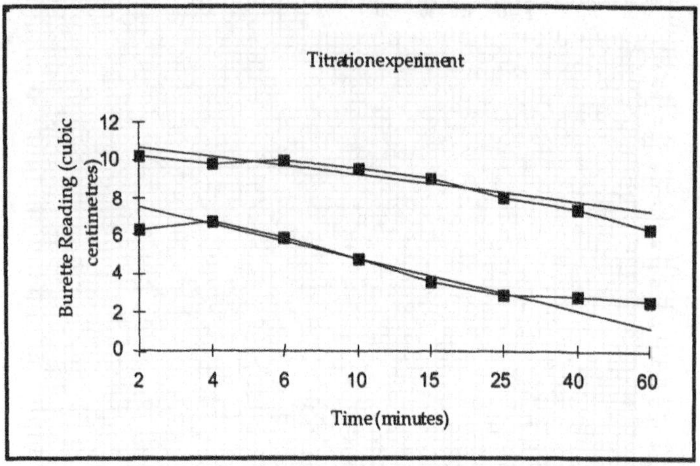

Figure 2.

Figure 2 highlights problems arising due to the student using computer software to graph data; note the non-linear scale on the horizontal axis. This is likely to be an increasingly frequent problem as students attempt to use technology to process data. The student's science teacher did not identify this lack of competence. Fortunately the gradients of the linear models shown on the graph were calculated and interpreted correctly.

The problem of the mathematical competence and confidence of staff involved in the delivery of these pre-vocational courses is one receiving much debate, at least within the mathematics education community. Attempting to contextualise much of the mathematics on these courses may be motivating for students but provides problems for teachers. Mathematics teachers are often not, by design, involved in the delivery of the course. In such instances we must rely on teachers within the vocational area to support the application of mathematics within the course; this can be fraught with problems where staff feel inadequately prepared for this. As is demonstrated here this practice can lead to problems even within contexts where one might expect subject-specific staff to have a relatively high facility with mathematics; in these instances we suspect that the focus of the teachers was on the science involved rather than the mathematics. On the other hand where mathematics teachers are involved in some delivery of *Application of Number* for students they often feel most comfortable when the mathematics is decontextualised, or at most is presented in an everyday context.

The two examples of graphs above were taken from students' work originating in the science units of the GNVQ. They illustrate how the general mathematical competence overarches the students' work in science; if the student is able to demonstrate competence with 'handling experimental data graphically' she will be helped to make sense of scientific phenomena that are new to her.

5. SPECIFYING NEW QUALIFICATIONS

At the same time as our work progressed on the development of *general mathematical competences* for science students it was recognised by Capey (1996) that the specification of pre-vocational courses was over-elaborate and leading to fragmentation of the curriculum and lack of coherence. A new model of specification for pre-vocational courses has therefore been developed that promotes coherence by demanding that students complete a small number of synoptic pieces of work that bring together clearly defined knowledge, skills and understanding. This mirrors the thinking underlying *general mathematical competences* where an aim has been to identify relatively substantial amounts of mathematics organised in ways that are useful to students.

Additionally concerns have been expressed about the relatively low mathematical demands of the majority of post-16 students' courses. In an attempt to encourage a greater number of students to follow a course in mathematics to a greater depth than that of the Key Skill, *Application of Number*, we have therefore developed, on behalf of the government's Qualifications and Curriculum Authority, new qualifications that are based on the application of mathematical principles. The specifications of the qualifications attempt therefore to reflect the principles underlying *general mathematical competences* allowing students to develop an area of application of mathematics to some depth which supports their extra-mathematical interests, studies or intended vocation.

It is the intention that these new qualifications will specify in general terms how students will be expected to demonstrate their application of mathematics allowing students to choose for themselves the contexts in which they will work. Units have been developed at three levels encompassing a wide range of ability that one can expect of students post-16. These will be piloted during 1998-2000.

Figure 3 shows a framework of unit titles at each of the three levels. A brief description of each unit can be found in Wake, 1997.

Each unit defines in a section 'Portfolio Evidence' the type of material that students must collect in a portfolio that will be assessed by their teachers. This should allow students such as those working as apprentice jewellery shop assistants, described by Corrine Hahn, or those training to work in the construction industry, described by Annie Bessot, to situate their demonstration of mathematical competence in contexts that are real and motivating to their other studies and/or work.

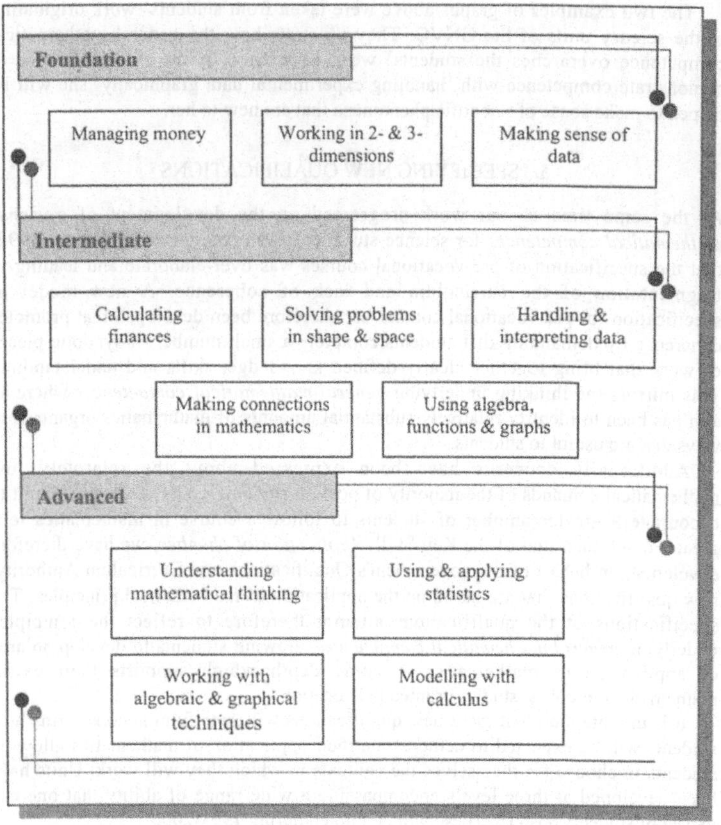

Figure 3.

Figure 4 exemplifies the form in which such evidence will need to be organised by students working on the Intermediate Level unit: Modelling *with Algebra, Functions and Graphs*.

What you need to produce

Reports of at least two investigations you carry out into situations in which you show your use of
a graphs,
b linear, proportional and non-linear models,
c algebra.

You should show evidence that in places you have used both estimation and checking to ensure that your work is accurate.

In the totality of your reports you should:

a

use graphs and graphical techniques to:

- plot
  - (i) data
  - (ii) at least one linear model
  - (iii) at least one other non-linear model

where at least one is plotted by hand and at least one is plotted using either a graphic calculator or computer software.

- explain how each graph relates to what is happening in the real world situation(s) you investigated by identifying and calculating any appropriate key features of the graphs including
  - (i) intercepts with axes
  - (ii) gradients
  - (iii) areas under graphs
- predict what will happen for cases for which you have no data
- solve problems involving
  - (i) linear, and
  - (ii) non-linear

  functions

b

find:

- one linear or proportional function to model data (using values you have found for the gradient of the line and its intercept on the vertical axis), and
- one non-linear function to model data

Figure 4.

It is the intention that students working in fields as diverse as, for example, Science and Business Studies can use these specifications and that their evidence will arise naturally from their studies in these areas. In addition to being assessed on the work that they produce for their portfolio, students will also have to pass a written examination. This will as far as possible, mirror the type of work that students will gather for their portfolio. Assessment by examination is being piloted that will allow students to work with data from Data Sheets they will have had the opportunity to familiarise themselves with in advance (c.f. Williams and Kitchen (1993), for instance).

Figure 5 shows such a Data Sheet from the sample examination of the Intermediate Level unit: *Modelling with Algebra, Functions and Graphs*. The data provided has been chosen as potentially having an appeal to students who are likely to study this particular unit, that is, those with both scientific/technical and business/commercial backgrounds.

Free-standing mathematics Unit, Intermediate Level: Using algebra, functions & graphs	**DATA SHEET This must be available to candidates for up to 14 days prior to the examination**

Body Mass index

Body Mass index, B, is used to decide wether or not your are overweight.

It is calculated using $B = \frac{m}{h^2}$, where m is your mass in kilograms and h your height in metres.

When you have calculated your Body Mass Index, B, you can use the table opposite to decide if you are under- or over- weight.

	Status
$B < 18$	**Under-weight**
$18 \leq B \leq 25$	**O.K.**
$25 \leq B \leq 30$	**Over-weight**
$30 \leq B \leq 40$	**fat**
$40 \leq B$	**very fat**

Converting

Height

If you know your height in feet and inches you can convert it to a height in metres:

1 convert feet to inches using:
1 foot = 12 inches
2 calculate your total height in inches
3 convert to metres using
1 inch = 0.0254 metres

Examples
Height: 5 feet 7 inches
1 5 feet = 5 x 12 inches
= 60 inches
2 5 feet 7 inches = 60+7 inches
= 67 inches
3 67 inches = 67 x 0.0254 metres
= 1.7018 metres
= 1.70 metres
(to the nearest centimetre)

Mass

If you know your mass in stones and pounds you can convert it to a mass in kilogrammes:

1 convert stones to pounds using:
1 stone = 14 pounds
2 calculate your total mass in pounds
3 convert to your mass m kilograms using
$m = 0{,}4536 \times p$, where p is your mass in pounds.

Examples
Mass: 9 stones 7 pounds
1 9 stones = 9 x 14 pounds
= 136 pounds
2 9 stones 7 pounds = 136+7 pounds
= 143 pounds
3 $m = 0{,}4536 \times p$
= 0,4536 x 143 kilogrammes
= 64.8648 kilogrammes
= 65 kilogrammes
(to the nearest kilogramme)

Figure 5.

Figure 6 shows the type of question that it is proposed students will be required to answer using the data in the written examination.

Section A **Use the Body Mass Index section of the Data Sheet.**

1

a A male student is 1.8 m tall.
Write an expression for his Body Mass Index , *B*, in terms of his mass, *m* kg.

b Over a period of time this student slims so that his mass reduces from 15 stone to 10 stone.

Complete the table below to give the student's:
(i) mass in kilograms, to the nearest 0.1 kg
(ii) Body Mass Index, to the nearest 0.1
for masses of 12 and 13 stones

Mass (stones)	15	14	13	12	11	10
Mass (kilograms) *to nearest 0.1 kilogram*	95 .3				69.9	63.5
Body Mass Index *to nearest 0.1.*	29.4				21.6	19.6

2 Plot, on the graph paper opposite, a graph of the student's Body Mass Index against mass (in stones) as he slims from 15 to 10 stone.
Plot
- mass on the horizontal axis
- Body Mass Index on the vertical axis

3 On your graph indicate the range of masses, in stones, between which the student can be considered to be "overweight".

4

a State whether your graph represents a
- proportional,
- other linear,
- quadratic, or
- other non-linear function

b Explain your choice of function

5 A ladies lightweight rowing team every member must weigh less than 65 kilograms.
Assume that a particular member of a crew weighs 65 kg.

a Write an expression for her Body Mass Index, *B*, in terms of her height, *h* metres

b Rearrange your expression to give *h* in terms of *B*.

Figure 6.

6. TAXONOMY OF LEARNING

An inevitable feature of the *general mathematical competence* is that it makes explicit to students ways in which mathematics can be organised; basic knowledge, skills and understanding are applied to solve real problems and interpret the real (for example, the scientific) world. It is the intention that the *general mathematical competence* does this rather than leaving to chance that students learn this by assimilating it as part of their experience. An emphasis of pre-vocational courses is that of applying knowledge. It is therefore an important facet of the qualification that needs to be taken into consideration when comparing it with others. In pioneering a methodology of comparing various qualifications at this level Coles and Matthews (1996) draw on the work of Bloom (1956) to identify a taxonomy of learning strategies. As part of ongoing research into the required and expected capabilities of students following pre-vocational and other courses, we have simplified Coles & Matthews' six categories of 'focus of performance' into four categories of '*focus of activity*'. This allows us to give a dimension in which we can describe how students are working with mathematics rather than simply defining content by type and increasing sophistication. Figure 7, below, attempts to illustrate this in the vertical axis: the horizontal plane allows one to categorise mathematics in the traditional way of a syllabus, that is, by domain such as algebra or functions, and by increasing sophistication (for example, in the algebra domain, by referring initially to linear functions and then progressing through quadratic and trigonometric functions etc.).

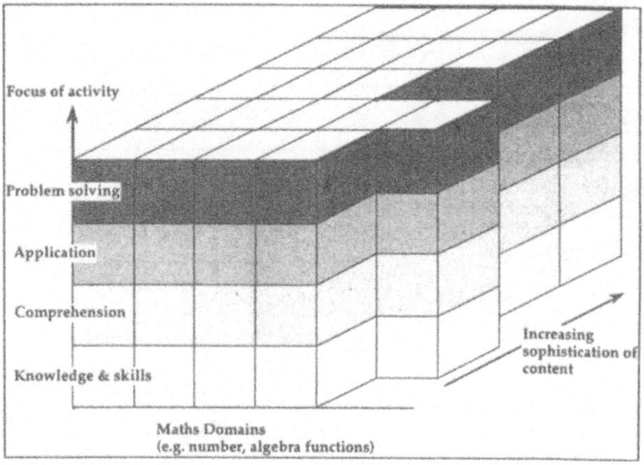

Figure 7.

The categories of 'focus of activity' used are:

Knowledge and skills

Students working at this level will be able to recall terminology, facts, conventions, methodology principles and generalisations. In terms of mathematics capability this will involve them in becoming technically fluent with a range of techniques at a basic level.

Examples of working at this level include:

The recollection of the convention that positive rotations are measured clockwise; the ability to factorise a quadratic expression...

Comprehension

Students working at this level will be involved in activity that involves translation from problem statements to symbolic notation, interpretation of mathematical information into understanding of real world behaviour, and the ability to apply mathematical abstractions when their use is specified.

Examples of working at this level include:

The translation of problems into symbolic notation; the ability to interpret the gradient of a line as a coefficient of interest in an experiment.

Application

Activity of this type will involve students in detecting and understanding the relationships that parts of a problem have on each other.

For example, given the problem of calculating the volume of an egg, the student will be able to draw on mathematical techniques such as electing to use a series of cylinders to model the egg, find and use relevant measurements, and calculate the sum of the volumes of the cylinders.

Problem Solving

This involves mathematical work of the highest order and involves Bloomian activity of Synthesis and Evaluation whereby students draw upon elements from many sources and put these together in a novel structure and judge the value of their solutions, perhaps with reference to other exemplary work of this type.

Initial indications suggest that students demonstrating competence with the defined *general mathematical competences* and in the newly developed qualifications will be working at the *Application* level in this taxonomy. This reflects the emphasis on mathematical modelling as outlined by Blum (1992) that underpins them. If students are able to reorganise the mathematics of the general mathematical competences to solve considerably complex problems then they may be working at the Problem Solving level. Activity of this type may involve the breaking down of *general mathematical competence* barriers and their reorganisation to solve unfamiliar problems.

Further research work using this taxonomy of learning in assessing the demands of the new qualification will take place as part of the evaluation of the pilot.

7. FURTHER RESEARCH QUESTIONS

The strategy described here gives an alternative method for specifying and teaching/learning mathematics for pre-vocational and other courses. The proposed *general mathematical competences,* take account of internal coherence to mathematical study, while also paying attention to how we expect students to be able to apply mathematics to help solve problems and make sense of the area in which they have chosen to base their studies. This latter focus, allows students' mathematics to be seen to have immediate applicability to their course of study, thus providing clear motivation to students who have not chosen to study mathematics for its own sake.

The following key questions need to be addressed in this development over the coming period:

-Can / to what extent can *general mathematical competences* be developed to

(i) sit comfortably *across* pre-vocational and other courses, in particular those outside of the 'technology cluster'?

(ii) support mathematical practice in vocational settings?

-Will the new qualifications under development support and enhance students' abilities to successfully apply mathematics

(i) in their chosen areas of study?

(ii) in workplace settings?

8. REFERENCES

Bloom, B. S. (1956). Taxonomy of Educational Objectives - Book 1 - Cognitive Domain. Michigan: Longman.

Blum, W. (1992). Applications and Modelling in Mathematics Teaching. In M. Niss, W. Blum and I. Huntley, (Eds.), Teaching Mathematical Modelling and Applications (pp. 10-29). Chichester, UK: Ellis-Horwood.

Capey, J. (1996). GNVQ Assessment Review. London: NCVQ.

Coles, M. and Matthews, A. (1996). Fitness for Purpose: A Means of Comparing Qualifications. London: NCVQ.

Dearing, R. (1996). Review of Qualifications for 16-19 Year-olds. London: SCAA.

Ernest, P. (1998). Mathematical Knowledge and Context. In A. Watson (Ed.), Situated Cognition and the Learning of Mathematics. Centre for Mathematics Education Research: University of Oxford.

Jessup, G. (1991). Outcomes: NVQs and the Emerging Model of Education and Training. London: Falmer.

Williams, J. S., and Kitchen, A. (1993). Implementing and Assessing Mathematical Modeling in the Academic 16-19 Curriculum. In T. Breiteig, I. Huntley and G. Kaiser-Messmer, (Ed.), Teaching and Learning Mathematics in Context (pp. 138-150). Chichester: Ellis Horwood.

Lord, K., Wake, G.C., and Willliams, J. S. (1995). Mathematics for Progression from Advanced GNVQs to Higher Education. Cheltenham: UCAS.

Sutherland, R., and Pozzi, S. (1995). The Changing Mathematical Background of Undergraduate Engineers: A Review of the Issues. London: The Engineering Council.

Wake, G., and Williams, J. (1995). Mathematical Modelling in Vocational Courses: The Case of GNVQ Science. Paper presented at ICTMA.

Wake, G. D. (1997). Improving Mathematics Provision for Post-16 Students, Teaching Mathematics and its Applications, 16(4), 200-206.

Wolf, A. (1994). GNVQs 1993-4: A National Survey Report. London: Further Education Unit, The Institute of Education, University of London and The Nuffield Foundation.

Wolf, A. (1997). GNVQs 1993-7: A National Survey Report. The Final Report of a Joint Project: The Evolution of GNVQs: Enrolment and Delivery Patterns and their Policy Implications. Bristol: Institute of Education, Further Education Development Agency and The Nuffield Foundation.

SECTION 4

RESEARCH METHODS
FOR MATHEMATICS AT WORK

SECTION 4

RESEARCH METHODS FOR MATHEMATICS AT WORK

PREFACE

ROBYN ZEVENBERGEN

The focus of Section Four is the methods and methodologies used to study workplace mathematics. The authors in this section draw on a range of methodological perspectives and tools in order to investigate various aspects of the mathematics used across a range of contexts. However, in order that researchers can investigate workplace mathematics a number of key considerations need to be addressed. In his book, *Mathematical Enculturation*, Bishop (1988) proposed two forms of mathematics. Since this time, further studies on how people undertake mathematics both within and beyond the school context have yielded findings that have caused the recognition of a number of three key forms of mathematics. These can be broadly classified as three distinct categories – formal mathematics such as that undertaken by mathematicians; school mathematics such as that taught in schools; and everyday mathematics (or ethnomathematics) such as that undertaken by individuals or groups of people in their daily practices. Most frequently, there is little transfer between the different forms. What such studies have been most powerful in demonstrating is that the mathematics taught in schools has little resonance with the daily experiences of most people. Rather, people develop contextualised and usually personally meaningful and effective strategies for resolving everyday problems that have little resemblance to those of the school-taught mathematics. In coming to study workplace mathematics, it is key that there is a recognition of the form/s of mathematics underpinning the study.

1. ASSUMPTIONS ABOUT MATHEMATICS IN THE WORKPLACE

When studying everyday practices in an attempt to understand how practitioners develop and use their unique forms of mathematics, there are two distinct approaches. In the first, and perhaps most dominant approach, the practices are often viewed through the eyes of school mathematics. In such studies, the researcher attempts to uncover the 'frozen mathematics' (Gerdes, 1986) in the

Bessot & Ridgway (eds.), Education for Mathematics in the Workplace, 183—187.

practices. For example, in their extensive study of workplace mathematics, Hogan and Morony (see Section 2) had teachers track workers in order to see how they mathematised within their workplace. This knowledge was used to support teachers' understanding of how various fields used mathematics practically, thereby helping teachers understand the importance and relevance of mathematics beyond the school context. In contrast, the other significant body of non-school mathematics seeks to show how the two forms of mathematics are very different from each other. This body of research has as its underpinning assumption that the two mathematics are of very different forms and as such seeks to understand the mathematics from the perspective of the participants rather than from a school-mathematics perspective.

2. METHODS FOR RESEARCH IN WORKPLACE MATHEMATICS

In studying workplace mathematics, there appear to be three main forms of mathematical contexts which are influenced by situational factors. These range from high levels of school mathematics through to contexts where there is minimal school mathematics. Magajana (1998) identified three forms of methods found in workplace mathematics – where there is a given formal method which is then applied; unexpected mathematical tasks which had an element of mathematics necessary for resolution; and tasks where there was no school mathematics involved in the tasks. Depending on the degree of mathematics involved in a context or workplace, then the demands can be substantially different and in many cases, the degree of competency in task resolution (Zevenbergen, 1998). Cases where there is a high level of school mathematics are typically those of engineers and architects and in most cases there is a high level of pre-requisite training which results in higher levels of competency. In contrast, there are cases where there are high levels of mathematics involved but low levels of competency and/or confidence as has been found consistently in nurses' calculations of drugs. In cases of intermittent or medium levels of school mathematics, there is often evidence of school mathematics but it has been modified to the context and hence a more practical focus. These studies are most frequently of workers in everyday occupations such as bankers, retailers, tradespeople and so on. In the third instance where there is little to no mathematics, the workplaces have been so mechanised and routinised that the tasks being undertaken require minimal independence of thought from the worker. Such instances include work in fast food outlets such as MacDonalds where producers outside the fast food outlet, for example, have undertaken all measuring.

The authors in this section, using a variety of techniques and methods undertake many of the issues discussed here. The techniques and methodologies used to study workplace mathematics are varied and influenced by the research problems and questions. Just as in other areas of mathematical research, there is not a singular method for collection of data nor analysis. The authors in this section have used a range of techniques to study various aspects of workplace mathematics.

With the technologising of many workplaces and professions, there has been a decline in technical skills and a renewal in the [purported] needs for basic skills. This is an international phenomenon. Many workplaces are now so influenced by technology, such as computers and other automated devices, that high level (and low level) skills are no longer needed. One only needs to consider the production line of fast food outlets through to car manufacturing; the tools used by surveyors in the field; or the computerised drawing tools used by architects, to observe the effects of technology on the skills needed by employees. However, employers are bemoaning the lack of literacy and numeracy skills in prospective employees. In the past, mathematics education has gained considerable status for being a discipline which can be transferred across contexts. With conservative governments in most Western countries, there has been a heavy emphasis on literacy and numeracy skills as these have been blamed as the cause for high youth unemployment. The study of workplace mathematics is significant in that it can demonstrate the unique forms of mathematics developed in workplaces and their synergies with formal school mathematics. As can be seen from the authors in this section, a variety of tools and approaches are most useful in studying how and to what extent school mathematics is used in workplaces

In his study of engineering apprentices, Ridgway identifies a decline in the basic numeracy skills of engineering apprentices through a survey format, but subsequent interviews and questionnaires indicated that such declines did not impact on the success rates of apprentices. This reflexivity between two methods allows for greater reliability within the data and analysis, but often yields a greater insight into the issue under investigation. In Ridgway's study, his dual methods confirm that the mathematics undertaken by engineering apprentices is quite different from that taught in schools. The data yielded in his study confirms that the apprentices used a range of informal and formal techniques for problem solving in the workshop, often incongruent with that taught in the schools, but far more effective in the context of the workplace. In some tasks there was a high degree of accuracy called for (such as measurement tasks) and a very high level of competency in completion of tasks, yet with little resonance with school mathematics. In coming to this conclusion, he is able to challenge the call for back-to-basics reforms and argue for a more balanced approach to teaching school mathematics.

In contrast to Ridgway's practical approach to researching workplaces, Mercier takes a key construct, 'a tool', as the focus of his chapter and then through a theoretical analysis of the 'tool' demonstrates how the original function of the tool (in these cases, calculations) has been disguised and a routinised procedure replaces the original function. Mercier's approach is theoretical and historical. Using the examples of sardine counting and dairy feed calculations, Mercier demonstrates how the historical foundations for the 'tools' have been lost and the process, which is now non-mathematical, has become a simple procedure. The complex calculations for dairy feed have been replaced by a table which farmers rely on to work out the proportions of food mix required for particular stock. He argues that the original calculations were complex and multivariant, but mathematical reasoning is integral to the tool and not the user. The use of theory and history are the main tools used by Mercier in unpacking the mathematical foundations to two contexts.

In the third chapter in this section, there is marriage between theory and practice. In her chapter, Zevenbergen discusses ethnographic methodology. She argues for the ethnographic methodology for studying workplace mathematics. This chapter is theoretical and draws on exemplars from studies using an ethnographic approach. Zevenbergen argues for the use of ethnography for studying workplace mathematics as it provides the tools and philosophy for understanding the workplace practices from the perspectives of the participants. She argues that ethnographic studies have the potential to enhance our understandings of how participants come to work and understand their contexts from their perspective rather than as the mathematics educator attempting to uncover the hidden mathematics of the context. The chapter draws on a range of studies to illustrate the value of the ethnographic approach as well as to demonstrate the tools used within this methodology. Issues associated with conducting research within workplaces are also discussed.

The final chapter in this section draws on a range of methods and contexts and seeks to identify the intersections. Bessot takes a reflexive approach to understanding how experts and novices work out problems on construction sites. The approach used by Bessot identifies how people can use a range of methods to calculate retaining walls on slopes and how the demands and responsibilities of the different types of workers influences how tasks are solved. She proposes that there are substantive differences between the calculations made on site and those within the formal construction course. Using a range of techniques including interviews and surveys, she found that errors were seen to be a normal component of the construction site and that where accuracy was needed, this was the domain of a particular group of people within the site. She undertakes a comparative approach to her study and seeks to find differences between the construction site and the formal construction course.

The authors in this section have drawn on a range of methods and approaches to understanding the role and place of mathematics in the workplace. The techniques are varied and yield different forms of data and analysis, all of which aid in the construction of a better understanding of how people use, modify or reject the use of school mathematics in non-school contexts. These studies are important in the field of mathematics education as they compel educators to consider one of the more dominant 'myths' in the field concerning the value and status of mathematics in the curriculum.

3. FUTURE RESEARCH

There are a number of key issues which need to be developed in the future research into workplace mathematics. The first is the philosophical approach being used when conducting research. Researchers need to consider the biases and worldviews with which they enter the field as this provides the lens through which they will collect and analyse the data. For researchers who seek to uncover the 'frozen mathematics', there is a greater likelihood that such approaches seek to legitimate the status of school mathematics. On contrast, research which seeks to understand the processes used in workplace contexts as being those of the participants and arising from the uniqueness of the context are more likely to challenge the status quo of school mathematics. The implications of these two divergent approaches are profound for the status of school mathematics and subsequent work to be undertaken within the formal curriculum of schools – whether primary, secondary, post-compulsory or training sectors.

The second major trend in research approaches is to recognise the degree to which school mathematics or formal mathematics permeates the workplace as this has implications for the study. Where there are higher degrees of school or formal mathematics, there is a need for participants to have higher degrees of mathematical competence and confidence. The strategies used by participants may be influenced by perceptions of self held by the participants and this will influence how they work within that context. In other situations, there may be little or no mathematics needed and hence a study of workplace mathematics may be misguided in such contexts.

4. REFERENCES

Bishop, A. J. (1988). Mathematical Enculturation: A Cultural Perspective on Mathematics Education. Dordrecht: Kluwer.

Gerdes, A. (1986). How to Recognize Hidden Geometrical Thinking: A Contribution to the Development of Anthropological Mathematics. For the Learning of Mathematics, 6(2), 10-12.

Magajana, Z. (1998). Formal and Informal Mathematical Methods in Work Settings. In A. Watson, (Ed.), Situated Cognition and the Learning of Mathematics (pp 59-70). Oxford, UK: Centre for Mathematics Education Research, University of Oxford.

Zevenbergen, R. (1998). Mathematical Saturation within Workplace Settings. In C. Kanes, M. Goos and E. Warren, (Eds.), Teaching Mathematics in New Times (Vol. 2, pp. 709-715). Gold Coast: Mathematics Education Research Group of Australasia.

CHAPTER 13

THE MATHEMATICAL NEEDS OF ENGINEERING APPRENTICES

JIM RIDGWAY

Abstract. A study was conducted into the mathematical needs of engineering apprentices, triggered by a decline in the basic number skills of applicants reported by employers. This study explored and confirmed the reported decline. The mathematical needs of apprentices were investigated in a number of ways, including an ethnographic study, interviews, and the exploration of the psychometrics of predicting apprentice performance. The ethnographic study revealed that mathematical challenges of engineering differ from the mathematics taught in school. In particular, great precision is required, applied to a variety of mathematical techniques; a good deal of practical problem solving is necessary, too. The psychometric study revealed that conventional measures of educational attainment had high predictive validity; a test created to sample the mathematical skills directly involved in engineering had low predictive validity. Therefore, high-level skills required for a successful educational career generalise to practical work, whilst the acquisition of mathematical technique does not. One can conclude that 'basic skills' are not a foundation but rather are a component of mathematical education. It follows that mathematics education should encourage the development of a broad range of skills; practising the deployment of skills in a range of contexts should be encouraged.

What sorts of mathematics should schools teach? The question has no simple answer; indeed, this question is the focus of many national debates, especially in countries with centralised curricula. A wide range of educational practices can be observed in any international survey (e.g. Howson, 1991; and many authors in Malone, Burkhardt, and Keitel, 1989), reflecting a wide range of implicit answers. More explicitly, the relationship between education, employment, productivity, and economic success has been explored by comparing different countries' educational practices and the impact on student attainment at work (e.g. Wagner, 1986; Green and Steedman, 1993).

In the UK, a good deal of debate has focused on the role of 'basic skills' in mathematics. One view is to assert that mathematics is an hierarchical subject, where later learning depends critically on earlier learning. If one adopts this view, then it is reasonable to insist that students perfect their technique at each lower level, before they progress to the next level. Further, if education is to satisfy the needs of employment (for both students and for potential employers), then some hierarchically organised work-related training should be provided, as one strand in

Bessot & Ridgway (eds.), Education for Mathematics in the Workplace, 189—197.

the educational opportunities available to students. This reasoning underpins much of the design of National Vocational Qualifications in the UK (e.g. Jessup, 1991). A complementary view is to see mathematical development as being heterarchically ordered - with some core components developing in parallel, in no logically related sequence. Such a view of traditional school mathematics can be found in Dowling and Noss (1990), and in the context of training for employment in Smithers (1993). This paper offers an empirical approach to testing the claim that a high level of competence in basic skills is essential to later learning. It is set in the context of industrial training, and so is relevant to much of the current heated debate there (e.g. Smithers, 1993).

A study (Ridgway and Passey, 1993) was conducted on behalf of the Preston Area Industry Education Liaison Group and Preston Area COMPACT, into a range of issues concerning the role of mathematics in local engineering industries. The work was provoked by a strong feeling amongst employers that the basic number skills of young people applying for apprenticeships were is steady decline, and that this decline in basic mathematical skills was having a direct effect upon the costs of training. Teachers were uncomfortable about the claims of declining student attainment, and their discomfort was enhanced by a feeling that control of the curriculum had moved in relatively recent history from the school to the State, via the introduction of a National Curriculum. The research, then, was targeted upon issues which addressed strongly held feelings, key issues in cognitive psychology and pedagogy, and which also seemed to have considerable economic and political importance. The research set out to explore a number of key issues, notably:

-the apparent decline in basic number skills;

-mathematical problems met during training;

-current practice regarding recruitment, selection, and training in engineering industries;

-and the implications for cognitive psychology and pedagogy.

Because of the multidimensionality of the issues to be addressed, and the number of stakeholders (such as employers, students, schools and society at large) the study used a variety of sources of evidence, which included:

-data from two major engineering employers which enabled the hypothesised decline in number skills to be investigated, along with the validity of a number of measures of academic attainment in predicting subsequent performance at work;

-survey data from members of the Group Training Association;

-questionnaire data and group interviews from three small samples of apprentices;

-a brief ethnomethodological study, where this author worked in an apprentice workshop making a bolt from an unformed cylinder of metal under the close guidance of an instructor; this study afforded the opportunity for detailed in-situ observations of the daily tasks of apprentices, and some informal discussions.

1. EXPLORING THE PROBLEM: ARE THE NUMBER SKILLS OF APPRENTICES IN DECLINE?

One major employer reported a decline in scores over successive years on a test designed to assess those mathematical skills of direct relevance to the work of apprentices. A problem with these data was that there was no base line against which to interpret the putative decline - a large decline in number skills could reflect social changes where high attaining students no longer choose to enter engineering as apprentices; there seemed little point analysing these data. A second major employer uses the Number Test to select apprentices. Essentially, this is a test of basic arithmetic. The same employer uses AH4, which is an intelligence test focused on verbal and numerical patterns and relationships. Data were available which allowed the scores of applicants to be tracked over successive years. The data showed that scores on the Number Test suffered a dramatic decline, while AH4 scores did not, over a six year period. See Ridgway and Passey (1995) for a full account of these data. This evidence provided clear evidence that scores on basic number skills declined dramatically over a six year period, as employers had claimed. This decline cannot be accounted for in terms of more general trends in the quality of applicants which one might associate with drifts in the relative popularity of engineering as a career, easier access to higher education, or other social phenomena. It can be concluded that, indeed, scores on basic skills in the applicant pool have declined over this period. Potentially, this raises a serious threat to manufacturing industries, if this group of basic skills is, as claimed by trainers, strongly related to subsequent apprentice performance.

Do Apprentices Fail?

Overall, trainers reported that apprentice and trainee failure rates were modest. In a questionnaire survey completed by members of the Group Training Association, the commonest causes of failure were judged to be a lack of interest and commitment. Specific skill deficits in mathematics were not highlighted. Interviews with apprentices themselves produced similar responses. Most companies in the survey claimed to experience no problems in the mathematics of apprentices and trainees.

2. WHAT ARE THE MATHEMATICAL CHALLENGES OF ENGINEERING?

The mathematical challenges of engineering were explored by questionnaire, and via an ethnographic study. The most frequent problems reported in a questionnaire to the Group Training Association were: lack of awareness and familiarity with Imperial measurements ('English units' in the USA); lack of number skills, even with addition and multiplication, and certainly those associated with fractions and decimals. This is consistent with a much larger survey reported by Harris (1991).

As part of this research, the author worked for a day as an apprentice in order to obtain direct experiences of the mathematical and other skills demanded, and in order to question trainers and apprentices in the work situation itself. Apprentices are required to switch fluently between imperial and metric systems - for example, they have to make engineering artifacts to an imperial specification on machines which are calibrated in metric units, and vice versa. Imperial units still dominate in the USA, and metric units are used throughout Europe, so the problem is unlikely to go away. A great deal of table look-up in books ('Zeus') takes place, for example to discover how to set a machine so that it will cut a metal cylinder of a particular diameter in such a way as to create a particular screw thread on its surface. Tables of logarithms were used extensively, because they gave the results of trigonometric calculations in minutes and seconds (required for machine setting) rather than as the decimal numbers produced by the calculators commonly available at the time of the study. The use of log tables in schools used to be commonplace, and (presumably) the skills acquired there generalised to the extensive table look-up required of apprentices. Such skills now have to be acquired at work.

Sometimes, the mathematics in industrial practice has unique characteristics, at odds with conventional mathematics. For example, in the workshop, I was asked to report a micrometer setting which read 0.3541. I *responded 'zero point three five four one'*. The instructor paused, then said *'OK, 354 plus one tenth'*. The traditional unit of measurement in UK engineering is a thousandth of an inch - the 'thou' - so these are often reported as units and fractions. When asked how to convert imperial measures to metric measures, I said *'multiply by 2.54'* - which is the everyday way to convert inches to centimetres. After another long pause, and my 'everyday' explanation, the instructor replied *'you mean 25.4....'* followed by *'centimetres? who uses centimetres? it is such a huge unit - what use would it be here?'*

The impressions from the apprentice workshop were of mathematics focused on: measurement to high levels of precision; expert use of geometry, requiring considerable skills; extensive use of look-up tables; an emphasis on perfect performance (being only 99% correct on each of a sequence of measures, table look-up, calculations, and dial setting, would probably guarantee that every piece would be substandard); a good deal of learning to apply procedures, so that in the end, they became automatic; and the embedding of mathematics in a broader scheme, where solving a mathematical problem is just part of solving the larger problem of creating a metal object to serve some specific function.

2.1. What are the Mathematical Challenges of School Mathematics?

The National Curriculum for mathematics offers a clearly defined mathematical entitlement to pupils which schools are obliged to provide. Basic number skills do figure large in this entitlement, but many of the topics identified as being directly relevant to engineering mathematics (such as geometry, table look-up and imperial-metric conversion) do not have a high profile. Under the National Curriculum, pupils are entitled to a broad education, and schools have less autonomy to respond to local needs. The mathematical tasks described earlier as characterising engineering are likely to receive less attention in schools than they do at present, unless teachers choose to use exemplars of general mathematical principles, such as ratio and proportion, which are directly relevant to engineering, e.g. imperial-metric conversions. This suggests an impasse - engineering industries require mathematical skills which schools are hard pressed to find time to teach.

2.2. *How Do Apprentices Perceive Mathematics at School and Work?*

Twenty eight apprentices completed a questionnaire to explore their feelings and beliefs about the role of mathematics at school, in training, and at work. After the questionnaire was completed individually, a broadly based discussion took place with the whole group.

-16 apprentices (4 no response) reported that they had enjoyed mathematics at school;

-18 apprentices (3 no response) reported that they felt confident about their mathematical skills when they left school;

-23 apprentices reported that they had to go over topics since they left school in order to relearn them;

-20 apprentices (5 no response) reported that they felt confident about mathematics at work.

The majority of apprentices in this sample had enjoyed school mathematics, and were confident in their mathematical skills when they left. Nevertheless, they felt that many topics needed to be relearned. Mathematics was seen to be more relevant to the job than either English or Physics, but no more important than a range of non-academic skills such as getting on well with people.

3. WHAT PREDICTS APPRENTICE PERFORMANCE?

One major employer uses a procedure based on the quality of every piece of work produced, to determine a score for each apprentice, at the end of their first year of training. This provides an almost ideal criterion measure of apprentice performance, against which different predictors can be judged. Here this will be called the Apprentice Score, and will be used to explore the predictive validity of different tests used, including measures of educational attainment, and measures of basic skills on entry to the industry.

One of the tests used for the selection of apprentices is a mathematics test which was designed collaboratively by trainers and by local mathematics teachers specifically to assess applicants' performance on mathematical tasks directly relevant to engineering apprenticeships. One might expect, therefore, that this test would be a good predictor of the Apprentice Score. It contains items such as:

-calculating the surface area of a cuboid, given its length, width, and height;

-calculating the length of a side of a right angled triangle;

-choosing the fraction nearest to a given decimal number;

-calculating the area of a metal sheet when a rectangle is cut from it.

Because of the timing of the start of the apprenticeships, and the announcement of the results of the national examinations which take place at the end of compulsory schooling (the General Certificate of Secondary Education (GCSE)), GCSE passes are not a condition of entry. So data are available for candidates with a spread of scores in different GCSE subjects.

Table 1 shows the predictive validity of a number of measures of educational attainment for apprentices recruited in 1989.

Selection Tests	*n*	*Apprentice Score*
Interview	111	0.248
Mathematics Test	109	0.067
GCSE Mathematics	102	0.388
GCSE Physics	94	0.501
GCSE English	101	0.339
GCSE Drawing	61	0.475
GCSE 'Practical'[1]	75	0.433
Total GCSEs[2]	102	0.448

Table 1. Correlations between different predictors and a summary score reflecting apprentice performance over a year

It is clear that scores on the mathematics test used for selection purposes are unrelated to apprentice performance; GCSE mathematics grades are quite strongly related, as are grades on other GCSE subjects and total passes at grade C or above. The mathematics test is used to reject just half of the applicants, so its failure to predict the end of year criterion cannot be explained in terms of a reduction in test score range.

These data show clearly that the mathematical skills which are most directly relevant to apprentice training are some broad based mathematical competencies, rather than a small core of key techniques. These data argue strongly against the case that basic skills are hard to acquire, and underpin mathematical performance in general. If that were so, the Mathematics Test would be strongly related to Apprentice Score. A fuller discussion of these data is provided by Ridgway and Passey (1995).

The data suggest that the decline in basic skills is unlikely to require changes in industrial training, or to pose a threat to manufacturing industry. High levels of technical skill are clearly essential to the work of apprentices, and should not be dismissed as unimportant, or downgraded. However, the data on the predictive validity of different sorts of mathematical knowledge suggest that appropriate skills can be learned during the apprenticeship, and need not be the basis for selection.

One can conclude that the advocacy of 'back to basics' is quite inappropriate - knowledge of applicants' basic skills predicts nothing about their subsequent job performance, whereas a measure of general mathematical competence proves to be a moderately good predictor.

4. MATHEMATICS AT SCHOOL VERSUS MATHEMATICS AT WORK - RESOLVING THE PARADOX

The mathematics taught at school, and the mathematics used in engineering differ in a number of important respects. Scores on tests intended to sample mathematics relevant to engineering show a significant decline over recent years, and indeed, the levels of performance on some individual items were surprisingly low. Paradoxically, apprentices expressed confidence in their mathematical skills at school and at work, and claimed (broadly) that they met mathematics relevant to work while at school. A test designed by trainers of apprentices and teachers to sample the mathematics of direct relevance to apprentice training failed to predict end of year performance; grades on GCSE mathematics, and several other school subjects, were good predictors of end of year performance. How can these paradoxes be resolved?

Employers' focus on the decline in technical skills, and their use of tests of mathematical technique for the purpose of selection reflects a view that perfection of these techniques is hard to acquire, or that they reflect some more general purpose mathematical abilities, or that they are some kind of foundation which must be firmly in place before thinking is allowed (as advocated by the 'back to basics' movement). The data on predictive validity give clear evidence against these ideas; so too do apprentices self reports.

Learning is dependent on context: a number of extensive studies (e.g. Lave, 1988; Carraher, 1991) have shown that students can carry out complex calculations in some circumstances (e.g. selling fruit; scoring ten-pin bowling; playing darts) that they cannot do in other settings (e.g. school). Learning mathematics in school, then applying it to a rather unfamiliar industrial context is likely to require a good deal of relearning. Relearning is acknowledged openly by the apprentices, but might be interpreted by others as ignorance.

Schoenfeld (1992) argues cogently that mathematical thinking is multidimensional, and that an account of a person's intellectual 'resources' (i.e. their access to facts and procedures) only tells part of the story. In any mathematical performance, a number of dimensions interact, which include strategies, control, beliefs, and metacognition. These dimensions are not structured hierarchically. The intention of GCSE was to shift the emphasis of school mathematics away from an exclusive focus on the acquisition of techniques, to include some learning of how and when these techniques can be applied, and to develop process skills of the sort described by Schoenfeld. The evidence on the decline in test scores could well be an indication of what has been lost; the high correlations between several GCSE grades and apprentice scores perhaps shows what has been gained.

In informal discussion, apprentices offered a sophisticated account of the nature of mathematics, and the nature of learning, which contrasts starkly with some of the beliefs implicit in some industrial selection practices, and with much of the debate about the appropriate composition of mathematical instruction. If students are to generalise their mathematical learning, they need to practice deploying what they have learned in a variety of contexts. In parallel, learning some aspects of mathematical technique to a very high automaticity (be it the multiplication table or the square root algorithm) and using this technique in situations which are personally relevant, also seems an essential ingredient of any mathematical education. Such mathematical activities need not precede other sorts of mathematical activity, however. Learning mathematical technique is likely to be easier if one can see a real purpose, and if feedback about accuracy is swift and personally relevant.

5. SOME CONCLUSIONS

Many of these conclusions are consistent with much that has been advocated in the mathematics education literature for some time (e.g. Collins, Brown, and Newman, 1986; Ralston, 1989) and in National Statements on mathematics education (e.g. Cockcroft, 1982; National Research Council, 1989):

-successful application of mathematical technique is essential in engineering (and elsewhere);

-the competencies learned from a broad based education generalise to practical work (apprentice performance, here); acquisition of mathematical technique does not;

-mathematical technique is not a 'foundation' but rather is a component of mathematics education;

-mathematics education should encourage the development of a broad range of skills, and some successful application of technique;

-deployment of skills in a range of contexts should be encouraged;

-the perfecting of mathematical technique should not be pursued for its own sake.

6. REFERENCES

Carraher, D. (1991). Mathematics In and Out of School: A Selective Review of Studies from Brazil. In M. Harris (Ed.), *Schools, Mathematics and Work*. Basingstoke: The Falmer Press.

Cockcroft, W. H. (1982). *Mathematics Counts: Report of the Committee of Enquiry*. London: HMSO.

Collins, A., Brown, J.S., and Newman, S. E. (1986). Cognitive Apprenticeship: Teaching the Craft of Reading, Writing and Mathematics. In L. B. Resnick (Ed.), *Cognition and Instruction: Issues and Agendas*. Hillsdale, NJ: Lawrence Erlbaum Associates.

Department of Education and Science and The Welsh Office. (1991). *Mathematics in the National Curriculum*. London: HMSO.

Dowling, P., and Noss, R. (1990). *Mathematics versus the National Curriculum*. London: The Falmer Press.

Green, A., and Steedman, H. (1993). *Educational Provision, Educational Attainment and the Needs of Industry: A Review of Research for Germany, France, Japan, the USA and Britain. Report Series No. 5*. London: National Institute of Economic and Social Research.

Harris, M. (1991). Looking for the Maths in Work. In M. Harris (Ed.), *Schools, Mathematics and Work*. Basingstoke: The Falmer Press.

Howson, G. (1991). *National Curricula in Mathematics*. Leicester: The Mathematical Association.

Jessup, G. (1991). *Outcomes: NVQs and the Emerging Model of Education and Training*. London: Falmer.

Lave, J. (1988). *Cognition in practice: Mind, Mathematics and Culture in Everyday Life*. Cambridge: Cambridge University Press.

Malone, J., Burkhardt, H., and Keitel, C. *The Mathematics Curriculum: Towards the Year 2000*. Perth: Curtin University of Technology.

National Research Council. (1989). *Everybody Counts: A Report to the Nation on the Future of Mathematics Education*. Washington DC: National Academy Press.

Ralston, A. (1989). A Framework for the School Mathematics Curriculum in 2000. In J. Malone, H. Burkhardt and C. Keitel, (Eds.), *The Mathematics Curriculum: Towards the Year 2000*. Perth: Curtin University of Technology.

Ridgway, J., and Passey, D. (1993). *Mathematics at Work*. STAC Monograph: Department of Psychology, University of Lancaster.

Ridgway, J., and Passey, D. (1995). When Basic Mathematics Skills Predict Nothing: Implications for Education and Training. *Educational Psychology, 15*(1), 35-44.

Schoenfeld, A. M. (1992). Learning to Think Mathematically: Problem Solving, Metacognition and Sense-making in Mathematics. In D. Grouwes (Ed.), *Handbook for Research on Mathematics Teaching and Learning*. New York: Macmillan.

Smithers, A. (1993). All Our Futures: Britain's Educational Revolution: Channel 4 Television, London.

Wagner, K. (1986). *Relation Between Education, Employment and Productivity and their Impact on Education and Labour Market Policies - A British-German Comparison*. Berlin: European Centre for the Development of Vocational Training.

7. NOTES

[1] 'Practical' refers to the best grade obtained on any practical GCSE subject taken, such as Craft, Design and Technology.

[2] Total GCSEs refers to passes at Grade C or above.

CHAPTER 14

IDENTIFICATION OF SOME MATHEMATICAL NEEDS LINKED TO THE USE OF MATHEMATICS AT WORK

AN EDUCATIONAL SURVEY[1]

ALAIN MERCIER

Abstract. This article offers a theoretical approach to a problem which is normally solved pragmatically without ever being raised at all: it attempts to define what mathematical techniques are necessary for tasks in which it is not evident that mathematics are being used. The concept of a tool introduced by Vygotsky helps to show how mathematical knowledge is inscribed within an artefact, so as to become invisible to its users. A study of the maths which are at work in occupational practice may thus be helpful in building up an understanding of the necessary relationship between the teaching of mathematics and the requirements of vocational training which, it is supposed, are fulfilled by that teaching. With the anthropological approach developed in maths teaching (mainly in France: Chevallard, 1991), it has been possible to test out a theoretical framework which is known for its application to vocational teaching problems. Thus, we emphasise the fact that a tool is socially determined and that this, in its turn, determines the way people think about problem-solving.

The purpose of this article is to investigate the way in which mathematical techniques are used as devices in occupational practice. The idea of a 'tool' (Vygotsky, 1930) is examined, to indicate both the device and its customary usage, with the intention of showing how mathematical knowledge becomes transparent to its users, just as a hammer became for a smith. The research is based on the anthropological approach developed in maths teaching (Chevallard, 1991) mainly in France. We are concerned here, then, with testing out a theoretical framework which is known for its approach to vocational teaching problems.

The mathematics known by those holding the old French 'school certificate' were sufficient for the needs of the great majority of the population until half a century ago; it was arithmetic for solving word problems. Most of the arithmetical techniques were founded on mechanical principles, written into systems whose workings were based on Archimedes' principle with occasionally a little practical chemistry or electricity thrown in. Today, however, the issue of vocational needs for mathematics is bound up with the background knowledge and techniques necessary

Bessot & Ridgway (eds.), Education for Mathematics in the Workplace, 199—208.

for the occupational use of mathematically complex technological devices. The example of a calculating device will be studied, together with its development, in order to support the work presented. In the past this device (the diagonal cross) was used by a range of technicians in agronomy, in laboratories as well as in cattle breeding, but it is now used only in teaching because in the world of stockmen, no one does calculations by hand anymore.

In order to identify a tool, we shall study all the 'actions which make up use' of a device. This will be a development of Mauss' theory (1936) that 'the primary and most natural technical object - and at the same time technical resource - known to man, is his body'. Following Vygotsky, we shall apply the term 'tool' to the combination of technical apparatus and the actions which this evokes (Mounoud, 1970), (Rabardel and Vérillon,1985). Finally, with Chevallard (1991), we shall emphasise the fact that a tool is socially determined: routine or traditional, it only becomes consolidated within an institution.

1. IS THERE A NEED FOR MATHEMATICS?

1.1. SAMPLE 1. The counting of sardines by Breton fishermen at the beginning of the century.

P. de Bonnault-Cornu and R. Cornu (1991) report an anthropological study into the counting practices of Breton fishermen. In the account below, the anthropologists' study has been reworked a little, in order to bring the underlying mathematics into a sharper focus.

In sardine canning factories until after the First World War, payment for sardines was made 'au grand mille' or 'millier grand compte' which corresponded to between 1,020 and 1,250 sardines, depending on which port of the Atlantic coast was involved. The principle goes back to 15th century working practices in the industry when this 'mille' or 'thousand' was the first link in the processing chain. The presses for drying sardines and herrings in barrels pressed two 'milles' into 'rondelles' or small rounds, and eight 'rondelles' made up a barrel. When asked about the counting of sardines by hand, to get to a total of a 'mille' which could be sold to the factory, old Breton fishermen described the following method.

Four baskets were used. One held the fish for the factory; threemore were used for counting. The counting unit was a handful of five sardines. For each handful of sardines put into the factory basket, the fisherman would put a token into the first counting basket (not surprisingly, this 'token' was a sardine!). For each handful of five tokens in the first counting basket, a token was placed in the second counting basket, and for each handful of tokens in the second counting basket, a token was placed in the third counting basket. So each token in the second counting basket corresponds to 25 sardines, and each token in the third counting basket corresponds to 125 sardines in the factory basket.

A factory basket was considered to be full when it contained 200 fish - five such baskets make up a 'mille'. The factory basket contains 200 fish when there is one token (125 fish) in the third counting basket and three tokens (3*25 fish) in the second counting basket and a handful corresponding to the token in the third basket. Five baskets made up in this way would thus have come to exactly one thousand, if the fishermen had not added the token fish to the factory basket. And in most ports they did, probably because these tokens were sardines, that those sardines had been counted but that it was not acceptable to start a new count with a basket that already contains 40 sardines. Most likely, it looked better to empty them before counting new baskets and this appeared to round off the operation!

The number of fish in a factory basket will then depend on the exact details of the counting strategy. If fishermen empty the baskets of token sardines as they go, they will have no more than 4 tokens in any counting basket at any time. When there are 200 fish in the factory basket, there will be no tokens in the first counting basket, 3 tokens in the second, and one token in the third. But should token sardines be thrown back into the uncounted sardines and be counted again? Given that fishermen simply add tokens to the factory basket in turn, when there are 200 counted sardines in the factory basket, there will be 40 tokens more from the first counting basket, 5 from the second (leaving 3 in the second, and one in the third). When all these tokens are added to the factory basket the 'grand mille' now contains (200+49)*5 = 249*5 = 1,245 sardines. Fishermen used to add one more in each factory basket 'pour le compte', and a barrel of 16 'milles' contains 20,000 sardines! The method has the advantage that large numbers of fish can be counted accurately, and that fishermen can stop counting mid-basket and continue later without error. So, with the device, counting by a group of people can be done without error. The pure base 5 method produces an accurate result; the counting practice avoids the need to recount the tokens. Moreover, no more than 4 tokens are leaving in the counting baskets and it is easy to make sure that no mistake has been made (accidentally or deliberately), by checking the contents of each counting basket.

This description serves as a model for a general study of the practice of mathematics in the working world. The system of baskets is *a counting device*, linked to the action of making up the handful of five sardines and choosing a token one. The tool is meaningful for the fisherman; but the fisherman himself cannot count up to a thousand without the baskets (enumerating every number from 1 to 1,000 spends half an hour, nobody can count half an hour without making mistakes). Like Marcel Mauss or rather, on this particular point, E. Hutchins (1995), I take the view that an occupational system of mathematical thinking is present in a set of actions governed by an apparatus which includes objects (in the one case, the system of the line of baskets), parts of the body (in the other, the system of hands which pick up the handful and the token fish), and a human organisation: *the occupational organisation of work does the computing*. There are no obvious mathematics and no longer any need for mathematics. The counting is inscribed within the tradition of the actions which reproduce it; this tradition is internalised by the workers as a tool. Here again as in most ethnocultural studies on mathematics in daily life (see Zevenbergen, in this section), mathematics are useless because people do not need theories about what they know how to do. However… In the early twentieth century, as soon as they know how to count a thousand, workers in the sardine canning factories went on strike to get paid for the exact number of sardines they processed!

1.2. SAMPLE 2. Present-day use of INRA[2] *tables by agricultural technicians for feeding dairy cows*

In France, over the last forty years, wherever technical agriculture was taught, from occupational high schools taking the BEPAexamination[3] to schools for agricultural engineers where students, after a course in Biology[4], are selected by competitive examination or, inbetween these two, when preparing pupils for the intermediate level of the BTSA[5], teachers of stockbreeding used to carry on them a small red book entitled 'Feeding Tables for Cattle, Sheep and Goats'. It was an Aide-Memoire which is a handy field or classroom adjunct to the reference work on the theory of feeding ruminants and the corresponding computational software published by the INRA. The use of the Aide-Memoire was a mark of professionalism in the stockman: both in early and further training, *it would seem that the ability to calculate a ration is an indicator of theoretical expertise in the job.* The principles are set out in the tables on pages 7 to 12 and the calculations in the tables on pages 13 to 23, and useful information on 714 types of vegetation (fresh, ensiled, dried; tubers, cereals, grain, etc.) in pages 24 to 115: for example on page 16, table 1.4, an 'example of rationing for a herd of adult cows inprocess of lactation' may be found; on page 19, table 1.7, 'examples of calculations of mineral supplements to basic everyday rations for dairy cows in production': four pages of examples, without a single generalised formula, and readers who are unaware of occupational techniques are still at a loss when it comes to doing the particular calculation they need.

A search for general directions for using the Aide-Memoire caused us to open a work on animal husbandry designed for highly qualified technicians[6]. But it only devotes a boxed piece of text in small lettering to the explanation of a 'general method', in which one learns that the problem relates to a system of several equations with several unknown factors, of which only the simplest case is set out (the coefficients are considered as constants): x and y being the quantities of fodder and concentrate measured in DM (dried materials), IC the ingestion capacity of the animal, BVF the bulk value of the fodder given (which depends on its humidity), BVC the bulk value of the associated concentrate (BVC = Sg*BVF, Sg being a substitution level), FUF the nutritional value of the fodder and FUC the nutritional value of the concentrate (measured in Fodder Unit for Milk production per kg of dried material, FUM/DM), and FUNS the recommended nutritional supplement (in FUM), which is a function of expected milk production. The equations are:

$$x.BVF + y.BVC = IC \qquad \text{and} \qquad x.FUF + y.FUC = FUNS$$

But FUNS = EN + I, where EN measures energy needs for the upkeep of the animal and the milk production envisaged by the cattle breeder, whilst I is a correction factor which takes into account the effect of a very rich feed caused by a concentrated element in the food: the bulk value of the concentrate diminishes the capacity for ingestion. And the difficulty is multiplied because Sg is, like I, a function of x and y...

One then finds a diagram showing families of graphs from which values can be interpolated to give possible solutions to the equations. However, these are unusable as they stand since they are not given as axes, being without a linear scale: they 'show the calculations' carried out by computers, but it is not possible to work with them because they do not show the results.

The questions thus remain: how are calculations made? What actions are effected and to what apparatus are they linked? Examples of results and tables of results cannot replace general calculations. Is there anyone doing these calculations by hand any longer? This does not seems to be the case. We must therefore identify the hand calculations.by ourselves. We found it in Frossard (1991), a book for stockbreeding teachers, and we must assume that the teachers are the only ones who still do calculations by hand, when demonstrating INRA software. As Ridgway noted in this section, the technicians no longer do such tasks because it is now calculated automatically.

2. A CALCULATING TOOL

2.1. Numbers arranged in the form of a diagonal cross serve to calculate proportions in mixtures

It would seem that these traditional practices (Martinand, 1984) for calculating the diet of dairy cows did not require any mathematical theory or conscious counting: no doubt the knowledge used sprang from traditional primary school training, which is that of any technician aged over fifty and of his parents. This training corresponds to the resolution of problems by stages, where the mathematical reasoning is built into the calculating tool, and not left to the user. A check is provided by knowledge of the context in which the problem is set, as in most 'problems of practical arithmetic'. When the investigation led in this direction, we were able to discover and show the workings of the technique behind the practice: it was the diagonal cross which figures in all primary school textbooks of the 1930's in School Certificate classes[7], and which was described recently by 'a team of teachers of maths and animal husbandry' who had met to analyse 'an arithmetical device which had to be taught to pupils studying for the agricultural BEP' (Frossard, 1991). Mathematically speaking, the cross represents a quasi-universal *barycentric calculating tool*; physicists calculate centres of gravity as statics, chemists calculate mixtures, economists use weighted averages, etc. It was at one time a tool as algorithmic as multiplication or division, but it disappeared from teaching with the reforms of the 1970's, and barycentric calculus is nowadays reserved for sixth form pupils (17 year-olds) specialising in mathematics whilst calculation of weighted averages no longer appears on the syllabus of any high school class, in France.

**110 kg of a feed at 0.84FUM were mixed with 50 kg
of a given feed, so that a mixture at 0.87FUM was obtained.
What is the FUM value of the unknown feed?**

110 kgof A	at 0.84	110(0.03/50)=0.066
M		0.87
50 kgof B	at ?	0.87-0.84=0.03

So that a feed B of FUM value 0.87 + 0.066 = 0.936 was mixed.

This is a diagonal cross for calculating mixtures

How does the diagonal cross for calculating mixtures work? It can be explained in this way : the excess of 0.066*50 compensates for the missing 0.030*110 to arrive at 0.87. Through the algorithmisation it allowed, the use of the diagonal cross as a simple 'device for calculating mixtures' made calculating a mixture as transparent as carrying out any other operation: a routine. Thus in the Aide Memoire, dealing with an example is the only thing that 'reminds' the future worker of the cross. But, as a tool for solving problems, the diagonal cross allows us today to see the mathematics encapsulated in the method.

The teachers who re-discovered it were resolving a teaching problem of animal husbandry, so that use of the cross is increasingly being taught once again. According to the authors, the document in which this tool is presented '[...] was devised to contribute to further training for teachers [...] there is no way in which it can be considered a manual [...]' it presents the results of a training and research exercise which aimed to deal with 'a long-standing quarrel between maths teachers and those teachers of animal husbandry who instruct their pupils in the use of the diagonal cross.' All of which indicates that the cross, in agricultural practice, was a device belonging not to mathematics, but to the mathematical work of the stockman: in the BEPA examination, it is sufficient to 'be able to fill in the form'.

The diagonal cross thus has a dual effect: *from an academic point of view,* it allows mathematical theory to be made visible in practice in actions carried out in occupational life (which means that mathematics can be taught) and at the same time *from a occupational point of view* it allows the mathematical knowledge available to the stockman to be put into practice (which means that mathematics can be used automatically without the user having to think about it). But by indentifying the mathematical knowledge needed to theorise the practice, our survey also points out the mathematical knowledge needed to equip the practical device. The arithmetic reasoning that necessarily underlies calculating 'in the head' is the only way of controlling the operations in the various cases. For instance, depending on wether the cattle breeder is attempting to determine the composition of the additive required for a given mixture or wether he is looking for a fodder with which a particular additive can be used, the calculations are not the same. Yet this type of arithmetical reasoning has not been taught since the 'new maths' reform in the seventies, it disappeared from the curriculum with practical arithmetic. And since then, 'mental arithmetic' is no longer used at school. From now, no one has what it takes to use a technique as complex as the diagonal cross! The '*artifact*' needed by this device is an academic 'basic knowledge'.

2.2. *Working knowledge and basic knowledge*

Let us return now to the Aide Memoire, in order to study its use in greater detail. The first table, entitled 1.1, gives the total alimentary needs (subsistence and production) in energy (FUM) nitrogen (DIP, Digestible Intestinal Proteins) and minerals (Ca and P) for adult dairy cows of 600 kg. Weights are given in kilograms. LW is the live weight of the animal, M its daily milk production, t the butyric level of the milk produced. Different needs are given according to the milk production observed. Formulae would summarise calculations whose results figure in table 1.1: they can easily be reconverted.

$$FUM = 1.4 + 0.006LW + 0.44\,(1 + 0.015\,(t - 0.040))\,M$$
$$DIP = 0.095 + 0.0005\,LW + 0.048\,(1 + 0.015\,(t - 0.040))\,M$$
$$P = 0.000045\,LW + 0.0017\,(1 + 0.015\,(t - 0.040))\,M$$
$$Ca = 0.00006\,LW + 0.0035\,(1 + 0.015\,(t - 0.040))\,M$$

Table 1.2 gives an evaluation of the milk production afforded by a singlefeed, with FUM energy value and known bulk. We learn from this that a feed which is richer in nitrite is consumed in less substantial quantities relatively speaking (we say that it contains more bulk) which partly limits the intervention of the stock breeder. But the actual calculation presented in the demonstration was made on the basis of data in this table for the case of a standard animal with the most common fodder: a formula dependent on a parameter (bulk) would have given us the information equally well, and it can be simply obtained from the preceding formulae.

Table 1.3 gives the values of variables from table 1.2, on the different types of fodder available for cows: it thus proposes a variation of the bulk parameter.

Lastly, table 1.4 gives a calculation of dietary requirements for cases where fodder is mixed, following an example step by step, with the commentary and verification essential to the technician of cattle breeding. The Aide Memoire does not go any further in describing calculations of feed rations for a dairy cow, but our observations of it show that teachers and agricultural engineers carry out the calculations set out in table 1.4 as a routine task. *This calculation - without it being specified - uses an arrangement which is precisely that of the diagonal cross. Only the answer is commented upon: one might just as well comment on a computer screen, while the technique by which the calculation was made does not even appear in the Aide Memoire.* As we expected, the basic knowledge at the origin of techniques used in occupational practices (here, mathematics) has become *transparent,* invisible when used professionally, so that the academic 'barycentric calculation' cannot come there, generate a technique of its own, and take the place left vacant. This phenomenon is quite similar to the one (about trigonometry) which Bessot points at, in this section.

Why, under these conditions, do the engineers of the INRA rely on a practice which they could have replaced by putting the problem as an equation and treating it with functional reasoning? Is it because the mathematical knowledge that is taught to students for the Scientific Baccalauréat or the BTSA would barely be adequate to make sense of these necessary mathematics? Is it because algebraic calculations must be carefully written down, whilst in the traditional treatment verbal reasoning or mental calculation are sufficient? Is it because 'the profession' would no longer recognise its traditional tools, which are supposed to be the symbol of professionalism? All these possible explanations reinforce one another without our being able to identify the key factor at work here.

2.3. *The ways of change*

But we are perhaps at a particular moment in history: technological change is making the old techniques obsolete and the feeding of cattle, sheep and goats is already a computerised technique, governed by theory which its users have no need to know. So that, from now, cattle breeding theories may change without any damage: stockmen need only change their software.

This point necessitates an additional enquiry. The INRA tables were drawn up by systematic experimental observation of the Institute's dairy herds. Certainly, the system of units used in analysing the observations constitutes an initial approach to calculating the rations of ruminants, and the practice of a succession of corrections using a series of tables gives a good approximation of an efficient ration. But although they explain actions based on sound reasoning, the tables do not describe the situations which necessitate use of these solutions. The 'laws' set out do not explain the phenomena on which the technician claims to act - as would, for example, a chemical or biological theory of digestion in ruminants. Although the notion of DIP ensures that the technician does not remain unaware of an essential phenomenon of ruminant-feeding, namely the production of animal proteins in the rumen (bacteria develop rapidly in the paunch, decomposing the plants ingested; the bacteria are digestible, and are afterwards digested in the intestine), we are dealing here with knowledge which has no disciplinary foundation (Vergnaud 1977; Artaud 1995). Knowing this allows a substantial economy of expensive supplements (soya cattle-cake, for example) to be made; however, the necessity of adjusting both FUM and DIP by the admixture of several food stuffs means that repeated *barycentric calculations* must be resorted to, which can only ever be approximate. Even if for teaching purposes it is still useful to study the problems to which knowledge is applied (see Ridgway, in this section), and in so doing analyse the mechanisms by which we think and act, the diagonal cross has lost - because of computers - its legitimacy.

Finally, the DIP model seems a poor one to researchers in animal feeding who are working to refine their approach, and calculate the ration of each animal according to its age, weight, physical condition and the comparative price of different supplements on the market. The tool represented by the person of the stockman is replaced by a calculator and software determined by a non-linear mathematical model, which it is sufficient to feed with information on demand. The needs of the profession for basic mathematical and biological knowledge have increased very rapidly, but satisfying them has become the prerogative of a more and more restricted number of researchers; these *knowledge are encapsulated in tools* where they have become less and less visible. This raises the questions of what form of training will be required for technicians whose activity will be governed by such software - where mathematics will no longer be visible; and what form of mathematical education (but also biological and chemistry education) will be necessary for technicians if they are to be able to follow the anticipated development of the tools they will be called upon to organise.

3. CONCLUSION

The analysis carried out and experience of the way in which the preceding developments have taken place, show that it is mainly the evolution in technology, where devices which are economical to use in mathematical terms are being introduced, which is producing the rapid change in social practices. The time has gone when this change was produced, from one generation to another, by increasing the common pool of knowledge which authorised the progressive emergence of new tools which were incorporated within it. Today, the old ways are being lost as the tools which gave rise to them are set aside (Guimpel, 1975), because the new devices no longer assume knowledge of tools.

Thus, in the present development of maths teaching for stockbreeding, the diagonal cross is replaced by 'families of graphs' to provide possible solutions to differential equations of a non-linear type. These show mathematics at work in the new types of models (Artaud,1995), without the need to set up occupational practices which are officially left to computers. They are of no use, but they bring evidence of greater understanding among top technicians and engineers. Work-sharing is gaining ground. With regard to scientific output, a few researchers in mathematics or the biology of ruminants and a handful of engineers in animal husbandry are alone in producing knowledge which can be used for the dietary needs of dairy cows (it is already happening with machines for counting fish and for calculating prices, change, etc.). Although schools of engineering offer a training in the basic knowledge necessary for stockbreeding skills, it would seem that only research engineers still practise mathematics for stockbreeding just as it is possible that they are now alone in practising biochemistry.

4. REFERENCES

Artaud, M. (1993). La Mathématisation en Économie Comme Problème Didactique, Une Étude Exploratoire. University of Aix-Marseilles II.

de Bonnault-Cornu, P., and Cornu, R. (1991). Savoir-faire, Savoir-mesurer. La Conserverie Nantaise. In Ministry of Culture and Communication (Ed.), Terrains 16, no spécial "Savoir-faire". Paris.

Chevallard, Y. (1995). La Transposition Didactique - du Savoir Savant au Savoir Enseignè, (2nd edition), Grenoble: La Pensée Sauvage.

Chevallard, Y. (1991). Dimension Instrumentale, Dimension Sémiotique de l'Activité Mathématique. Paper presented at the Seminaire de Didactique des Mathematiques et de l'Informatique 1991-1992, LSD2-IMAG Laboratory, Grenoble.

Guimpel, J. (1975). La Révolution Industrielle au Moyen-Age. Paris, Seuil: Translated from the English, Collection Points Histoire.

Hutchins, E. (1995). Cognition in the Wild. Cambridge, Massachusetts: The MIT Press.

Martinand, J. L. (1984). La Référence et le Possible dans les Activités Scientifiques Scolaires. Recherches en Didactique de la Physique, Actes du Premier Atelier International (CNRS).

Mauss, M. (1936). Anthropologie et Sociologie. Paris: Presses Universitaires de France.

Moumoud, P. (1970). La Structuration de l'Instrument Chez l'Enfant. Neuchâtel: Delachaux et Niestlé.

Rabardel, P., and Verillon, P. (1985). Relations aux Objets et Dévelopement Cognitif. Paper presented at the Actes des Septièmes Journées Internationales sur l'Education Scientifique, Paris: LIRESP and University of Paris VII.

Schlanger, N. (1991). Le Fait Technique Total. La Raison Pratique et les Raisons de la Pratique dans l'Oeuvre de Marcel Mauss. Terrain, 16, no spécial "Savoir-faire", 114-130.

Vergnaud, G. (1977). Activité et Connaissance Opératoire. Bulletin de l'A.P.M., 307(2), 52-65.

Vygotsky, L. (French translation by Catherine Haus). (1930). The Instrumental Method in Psychology. In B. Schneuwly and J.P. Bronckart, (Eds.), Vygotsky Aujourd'hui (pp. 39-47): Neuchatel, Delachaux, Niestle.

Works which are the subject of the survey

Frossard, G. (1991). *Mathématique et Techniques (Les Mélanges), Document INRAP 101*. Dijon: Ministry of Agriculture and Forests.

Gadoud, R., Joseph, M.M., Jussiau, R., Lisberney, M.J., Mangeol, B., Montmeas, L., and Tarrit, A. (1992). *Nutrition et Alimentation des Animaux d'Élevage, 2*. Paris: Foucher.

INRA. (1977). *Tables de l'Alimentation des Bovins, Ovins et Caprins*. Paris: INRA.

5. NOTES

[1] The example which forms the main content of this contribution was published in French by the journal DIDASKALIA, 4, in September 1994, under the title *Des études didactiques pourraient-elles aider l'enseignement des savoirs professionnels?* We wish to thank Laval University, the INRP and the publishers for their authorisation to use this work

[2] Institut National de la Recherche Agronomique (National Institute for Agronomic Research).

[3] Brevet d'Enseignement Professionnel Agricole (Vocational Teaching Certificate in Agriculture) level V.

[4] This teaching follows the general studies baccalauréat specialising in science.

[5] Brevet de Technicien Supérieur (Higher Vocational Training Certificate in Agriculture), a professional diploma (level III) which follows the baccalauréat in technology specialising in agriculture.

[6] R. Gadoud, M-M. Joseph, R. Jussiau, M-J. Lisberney, B. Mangeol, L.Montméas, A. Tarrit (1992).

[7] In the France of the time, the primary school certificate symbolised successful completion of compulsory schooling (at age 14).

CHAPTER 15

ETHNOGRAPHY AND THE SITUATEDNESS OF WORKPLACE NUMERACY

ROBYN ZEVENBERGEN

Abstract. This chapter examines the potential of ethnography and ethnographic tools for the study of workplace numeracy. It is proposed that ethnography is ideally situated for the study of practices within a workplace setting which give rise to unique forms of mathematics. By undertaking a study of such practices, narrow definitions of mathematics constrained by school mathematics are challenged. In so doing, richer understandings of mathematics can be developed which challenge current orthodoxies within the field of mathematics education. The chapter undertakes two key functions. The first is to provide a rationale for the use of ethnography in the study of the dialectic of context and mathematical understandings. The second is to recognise the unique aspects of conducting ethnographies within workplace settings where the purpose of the study is related to numeracy.

Over the past decade or so, the dominant discourses informing mathematics education have come under challenge from a variety of theoretical perspectives and their informing modes of criticism. In particular, the discourses that supported notions of objectivity; transferability of knowledge from one context to another; and absolutist notions of mathematical knowledge have been particularly vulnerable to these new challenges. Within these challenges, the role and recognition of the dialectic of context and mathematics has been particularly powerful. The contribution made by situated learning has been particularly illuminating in challenging the status quo. Further recognition of the power of the context on the construction of mathematical meaning and understanding has come from the discourses on constructivism. Together, these challenges have recognised that the context within which mathematical meaning is constructed is integral to the mathematics developed. Within such a position, it becomes paramount to study the dialectical relationship between mathematics and the context within which it is being used and developed. One means of undertaking such study is through the use of ethnographic techniques so that the role of context in developing mathematics can be better understood.

The value of ethnography as a tool for developing understandings of workplace knowledge lies in its fundamental assumption that knowledge and practices are developed within a social context. For example, Denzin (1978) argues that objects do not have any intrinsic meaning, but rather, their meaning is a result of human

Bessot & Ridgway (eds.), Education for Mathematics in the Workplace, 209—224.

actions towards them. More specifically, within the context of mathematics education, Eisenhart (1988) argues that 'all human activity is fundamentally a social and meaning-making experience' (p.102). Hence the fundamental assumption underpinning this chapter is that mathematical knowledge is seen to be the result of a human interactions within a particular social or discursive context. To this end, this chapter explores the use of the ethnographic methodology and methods as a tool for understanding the situatedness of mathematics and its relationship with the culture of workplace settings.

Within the ethnographic tradition, mathematical knowledge is seen as a consequence of particular human actions and the resultant interpretation of those actions. In order to develop an understanding of the mathematics undertaken within a workplace context, ethnography, in conjunction with the tenets of interpretivism, permits the interpretation of workers' actions and practices in relation to mathematics. Arguably the most significant contribution made by ethnography in the area of mathematics education is understanding the situation from the perspective of the participant. In the following sections, I present a rationale for undertaking ethnographic research, document the key assumptions underpinning the ethnographic approach, discuss the process and tools used within an ethnographic study and conclude with the implications of the outcomes of ethnographic studies for the field of mathematics education.

1. ETHNOGRAPHY AND THE STUDY OF MATHEMATICAL CULTURES

When undertaking cultural studies in a particular setting, one of the main obstacles facing the researcher is her/his own biases. At the level of culture, we are often unaware of them and hence while culture provides the lens through which we see and interpret the world, it simultaneously provides a bias. Barton (1997) suggests that when studying cultures of mathematics, there appear to be two main thrusts in the literature - one which looks at mathematics through the lens of western mathematics (such as that of Bishop's 1988 "Mathematics"), the other, through the eyes of the participants. The first of these approaches has defined mathematical thinking within a western mathematical framework and implicitly assigned status to a particular form of mathematical knowing thereby legitimating the power and status of western mathematics. Other approaches (such as that of Cooke, 1990; Watson, 1988) have recognised the unique systems for "encoding, interpreting and organising the patterns and relationships emerging from the human experience of physical and social phenomenon (Cooke, 1990, p. 5). In adopting this latter approach, Lave (1988) has argued for a process which is not constrained by Mathematics, but one which recognises and makes visible the socio-cultural aspects of mathematics. In recognising these two approaches, the literature on workplace mathematics can be seen to fall into these two broad categories. Accordingly, it is important to recognise the assumptions guiding the research. While ethnography, in its purest form, seeks to fall into the latter category, there are ethnographies which have made a contribution to the literature which fall into the former by seeking to document workplace settings with a western mathematics framework. It is not my purpose in this chapter to provide a critique of these two perspectives, rather to provide an argument for the use of ethnographic techniques in the study of workplace mathematics but to make explicit the recognition of these two perspectives and their impact on the study of workplace mathematics.

2. A RATIONALE FOR USING ETHNOGRAPHY

In this section I discuss the importance of the interaction of culture and cognition. The main purpose of ethnography is to represent the culture of a given community from the perspective of the participants. Where a community can be seen within social and cultural frameworks, this grouping can be extended to include workplaces. If one considers the differences between the workplace of banks, construction sites, fast food outlets and hospitals, it becomes quite apparent that there are substantially different social and cultural norms within each of these sites. The ethnographic tradition seeks to identify the unique cultural features of such sites. For the purposes of this chapter, such cultural features would include the mathematics embedded within these sites. In such rationale, mathematics is considered within a broad definition to include not only the formal mathematics synonymous with school settings, but also the unique processes and artifacts used within a particular context to solve the problems posed by the unique features of the workplace.

Until relatively recent times, the role of context has been largely ignored, but increasingly it has been recognised to play an integral role in the mathematics developed and used within that context. The trend in contemporary research has evidenced a change in emphasis away from performance to understanding and thinking (Mitchelmore, 1996)[1]. The development of theories and models of mathematical understanding increasingly recognise the role of context. Some large, comparative studies, such as TIMMS, have explicitly attempted to incorporate the role of context in analysing the outcomes of international performances (Robitaille & Nicol, 1994; Lokan, Ford, & Greenwood 1996/7a; Lokan, Ford, & Greenwood, 1996/7b). The ethnographic approach is most amenable to the documentation of this contextual aspect of mathematical understanding.

Further support for adopting an ethnographic approach to studying workplace numeracy can be found in mathematics education literature. Research undertaken in the areas of situated learning, ethnomathematics and constructivism have been particularly useful in documenting the impact of context on cognition, and hence mathematics learning and understanding.

2.1. *Situated learning*

Of particular importance to this chapter, is the contribution made by studies of situated learning and cognition where such studies have encompassed young school students in non-school contexts (see Nunes 1992, for a summary of this work) everyday activities such as shopping and weight watchers (Lave, 1984) candy sellers (Saxe, 1988); through to workplace settings (Masingila, 1993; Millroy, 1992;

Zevenbergen, 1996). These studies have demonstrated the highly contextual skills, knowledge and processes used in the resolution of everyday situations. This research has provided valuable empirical data confirming participants in these contexts develop and apply innovative and creative strategies in order to solve successfully the problems that they encounter. Such strategies are developed as a consequence of the goals, motivations and purposes of the immediate contexts in which the participants are located, thereby challenging the monopoly of school mathematics with its rigid procedures, processes and skills

Extending beyond contextual features, Lave and Wenger's (1991) research also documents the roles of people within these contexts and how transfer of knowledge is made possible. Their work with 'novices and experts' is of particular relevance to the study of workplaces. Unlike the formal school setting where transfer of knowledge is made possible through the explicit teaching-learning process, the workplace setting is often one of 'legitimate peripheral participation' in which the novice observes the expert until such time as the teacher or student feels confident or competent enough to try the skill being observed. This is seen as legitimate learning. Of further importance, the concepts of 'experts and novices' provide a useful methodological construct for conducting ethnographic research. In many cases, the experts have been immersed in their field for so long, they have embodied the practices of their profession and hence are unable to explain their actions. Just as culture is unobservable to the enculturated, so too, the processes and knowledge embedded within a workplace are often unspoken and hence unknown. In contrast, the study of novices reveals much of the culture of the workplace, and the mathematics involved in the situation.

The importance of Lave and Wenger's constructs of experts and the non-recognition of cultural components of the workplace were apparent in my study of bakers. The bakers did not measure any ingredients for a batch of 250 cookies, yet were able to produce a batch which were correctly mixed, accurate in number, and were of a consistent weight for each cookie. When estimating the size and weight of individual cookies, the baker was able to flip off a portion of the large dough onto the baking tray. Individual cookies were of similar size and weight and when packed into their cartons were of the correct overall weight to comply with the supermarket packaging demands. Bakers could only comment that they 'knew what the mix should look like'. Little formal measurement or teaching, as defined by school mathematics, occurs in this setting, yet the master baker and his apprentice were highly competent in their tasks – when the task is judged on performance as demanded in this context.

These studies, and their incumbent use of ethnography, have been instrumental in challenging the myths of transferability of mathematical knowledge from the school context into other non-school contexts, thereby challenging notions of key competencies of mathematics.

2.2. *Ethnomathematics*

The approach of ethnomathematics has also made a substantial contribution when considering broader cultural contexts. D'Ambrosio (1985) proposed that this area of research makes an important contribution in its recognition of the different forms of mathematics arising from different cultural contexts. He (D'Ambrosio, 1985, 1987) champions the recognition of a wider conceptualisation of mathematics to incorporate, and hence legitimate, non-academic practices. His research program seeks to identify culturally diverse examples of practices for their underlying modes of thought and patterns of reasoning. While there is considerable debate as to what constitutes 'ethnomathematics'[2], for the purposes of this chapter, the main point to be recognised is that it challenges the narrow conceptualisation of what is mathematics. One consequence of these challenges has been to recognise the different strategies and processes used within various social and cultural settings. In so doing, ethnomathematics highlights the need to reformulate mathematics from a position which assumed and supported notions of decontextualised and hence, transposable, knowledge and practices to one which recognised the highly contextualised (and cultural) practices of mathematics.

2.3. *Constructivism*

The literature on constructivism has played a critical role in the recognition of individual construction of meaning and the importance of interaction in the development of meaning and understanding. Of particular relevance to the study of workplace numeracy is the position advocated within radical constructivism. Von Glasersfeld's (1984) framework for radical constructivism allows for substantially different representations of knowledge to co-exist. In contrast to the realist position of a true representation of the world, a radical constructivist position supports the subjective reality informed by interactions with the social and physical environments. The constructivist literature has been powerful in recognising and documenting the impact of social conditions on the individual construction of mathematical meaning. Millroy (1992) recognises the synergies between constructivism and ethnography when she claims that:

> Constructivism encourages and acknowledges the legitimacy of student's own inventions and explanations, fosters acceptance of multiple representations of ideas, and emphasises the importance of actions and reflection. (p. 9).

By and large, this literature has been predominantly focused on school settings. Increasingly there is a growing recognition of the importance of documenting the mathematical understanding in non-school contexts.

The contributions made by these studies indicate the mutual relationship between culture and cognition. Culture provides the means through which members of a group construct meaning and mental representations. Mathematics is a part of this process. Through language, interactions and cultural artifacts, mathematical thinking is developed. The focus of this chapter is the use of ethnography as a tool for understanding the situatedness of workplace mathematics. Integral to the approach adopted in this chapter is the underlying assumption that mathematics is a socially constructed practice and knowledge. Eisenhart (1988, p.100) proposes that the research question posed within ethnographic approaches resembles 'why is mathematics teaching and learning occurring in this way in this setting?' Through examination of how mathematics is constructed within a particular workplace setting, it becomes apparent that different forms of mathematics are developed *in situ* and often these forms of mathematics and mathematical practices are far more appropriate and effective than the narrowly-defined mathematics of school-based mathematics. In order to develop an understanding of how participants construct their mathematical ways of knowing and understanding, ethnography offers the means by which interactions, behaviours and artifacts can be studied.

2.4. Key Assumptions of Ethnography

In summarising the general ethnographic approach, Glesne & Peshkin (1992) propose that ethnography is

> the anthropological tradition of long-term immersion in the field in with the researcher collects data primarily by participant observation and interviewing. ...[It should] be open to learning about social phenomena from a variety of perspectives and may, therefore, elect to mix techniques associated with different orientations (p. 10).

Within the ethnographic tradition there are a number of forms of ethnographies undertaken, each considered with dubious authenticity depending on the critics. The most rigorous form of ethnography is that which is conducted within the parameters of anthropology, involving sustained periods in the field, the researcher becoming a participant in the culture, provision of an account of the culture of the group. Ball (1988) uses the terms 'hard-core' and 'soft-core' to refer to the more pure or modified approaches to participant observation. These terms can be similarly applied to ethnographic approaches. The approach advocated by Glesne and Peshkin above indicates the hard-core approach and represents the ideal form of ethnography. Similarly, these approaches are those that Barton (1997) referred to as studying mathematics from the world view of the participants. Within the study of workplace numeracy, the approach adopted by Millroy (1992) is an excellent example of this approach. In Millroy's study, she became an apprentice carpenter in order to develop the mathematising of carpenters. Such approaches, while embracing the fundamental tenets of hard-core ethnography are highly valuable in documenting the mathematics developed and used within particular contexts, are less amenable to contemporary research contexts where such funding is highly competitive and restrictive. Where competition for grants is highly competitive and often politically driven the potential for long-term, labour-intensive research programs such a hard-core ethnographies of workplace mathematics can be restricted.

In contrast, there are those more generic approaches that adopt the principles of ethnography in seeking to develop a vivid description of a social or cultural group.

Some researchers (Wolcott, 1987; Rist, 1980) are openly critical of forms of ethnographies which do adhere to the philosophical and theoretical roots of the tradition or are mere descriptive reports, not embracing the concept of culture completely. Hard-core ethnographers are unlikely to consider such approaches as ethnographies. Rather, they are more congruous with the critical and postmodern ethnographies being developed in contemporary research literature.

The soft-core approaches are those which adopt many of the features of traditional ethnographic research but fail to adopt one or more of the key features of hard-core ethnographies. Those studies which Barton (1997) referred to as working with a Mathematics framework could be conceived as belonging to this category since they are constrained by the researcher's western mathematical world view. Without the sustained immersion in the field, there is less chance of documenting (or developing a restricted view of) the mathematics from the perspective of the participant, particularly when such an agenda is not in the consciousness of the researcher. In contrast, those researchers who are explicit about recognising (and documenting) the mathematics developed and used within a workplace from the perspective of the participants may be more akin to the fundamental goals of the hard-core ethnography but are unable to conduct the sustained immersion in the field.

While there are recognisable differences between hard-core and soft-core ethnographies, there are features that delineate ethnographies from other methodological approaches. These feature include:

1. Studying (and representing) the culture of a given social or cultural group

Ethnography arises from cultural anthropology and has as its key purpose to understand culture. Key ethnographers propose various interpretations of what constitutes ethnography but all resonate with the notion that it is the study of culture that is integral to the ethnographic purpose. For example, Wolcott (1987) argues that the purpose of ethnography is to describe and interpret cultural behaviour. Similarly, Spindler and Spindler (1987) propose that the purpose of ethnography is to make a coherent record of community behaviour and the participants' explanation of that behaviour.

2. The representation is that of the participants - not the researcher.

Unlike other forms of research, in ethnography, the authoritative voice is of the participants and not the researcher so that the accounts that develop from the research are from the perspective of the participants and not the researcher. In the study of workplace mathematics, this is significant in that the researcher must attempt to overcome her/his mathematical world view and take the world view of the participants. As has been documented in many of ethnomathematics studies, participants often do not see themselves mathematising. Rather, it is the researcher who sees the mathematics and often, that mathematics is rigidly defined within the school mathematics form.

3. Develops a theory, not confirms a hypothesis

From the work of Glaser and Strauss (1967), Millroy (1992) argues that the purpose of ethnography is not the confirmation of a hypothesis, but rather the development of a theory. Using notions of grounded theory, she argues for the "discovery of theory from data which has been systematically obtained through social research" (p. 70). Through the concurrent procedures of theory building, coding and data analysis, important aspects are discovered and noted for further observation. Thus the process of research and data analysis is continually developing throughout the fieldwork phase.

However, the above statements are not to be construed as absolute truths. As has already been noted, there are differences in the ways in which the ethnography is conceived due to the degree of cultural bias the researcher brings to the setting. Furthermore, the literature coming from a post-modern perspective challenges many of the assumptions listed above. The particular concern of how the researcher and researched impact on each other is worthy of consideration. Ellen (1984) argues

> The current view is more of one in which field "experiences" are "transformed" into data through encounters between researcher and researched; they are "translated" from one cultural context to another; and they are "constructed" drawing upon the personal and intersubjectivities of those involved (Ellen, 1984, p. 10)

However, notwithstanding these criticisms, ethnography as a technique for developing and documenting the culture of workplaces with their incumbent mathematics, has potential to contribute substantively to the field of mathematics due to its ability to highlight the different mathematics used in various contexts.

3. CONDUCTING WORKPLACE ETHNOGRAPHIES

There is an extensive literature on the conducting of ethnographic research. It is not my intention in this section to review this substantive work, but rather to provide a brief overview of the key tools used in conducting ethnographic research. In so doing, I alert potential researchers to the unique features which may be encountered when conducting ethnographies within workplace settings and which must be considered before access can be negotiated.

Negotiating Access

Within any ethnographic research project, negotiating access can be a complicated and complex process. Glesne and Peshkin (1992, p. 33) discuss the role of gatekeepers who 'must give their consent before you enter a research setting'. In the case of workplace settings, this can be the secretary who allows you access to the employer, it may be the site or store manager, the union delegates and so on. It may be further complicated due to the hierarchical structures within a workplace setting so that multiple gatekeepers are confronted in order to gain access to the site. Often an 'informant' is beneficial as this person can provide key information into the nature and structures of the workplace setting so that important and key personnel can be targeted rather than the being detained by an officious but relatively insignificant player.

Once the gatekeeper has been contacted and the negotiation process has commenced, it is important to have a set of clearly articulated research questions that can be understood by the gatekeepers and participants. The detail of these questions is debated in the literature as to how much should be revealed and the potential ethical dilemmas of revealing too much or too little (see Zevenbergen, 1998). At this initial meeting, it is also important to have a clear idea of what will be offered to the participants (for example, copies of final reports, rights of the researcher and researched, your expectations) and what might be potential problem areas in the research. For many, the research process is new and unclear, so it is the responsibility of the researcher to make as clear as possible, the potential for informed consent.

Within the workplace setting, there are a number of clearly identified areas of concern that are unique to this context and research. Particular issues confronting potential researchers are associated with the workplace health and safety issues and the notion of payment for work. These are often union issues associated with workplace employment and safety and their importance can not be ignored. As a novice entering the workplace, there are many safety issues that may need to be considered. In many cases, employers are not keen to have inexperienced 'workers' on site and are unwilling to take the legal responsibility attached to having a person on site. Millroy (1992) documents many of these problems she encountered with the negotiation of access to carpentry workshops.

A further issue compounding the negotiation of access to a worksite is payment for work. Where full access is sought and the researcher is seeking to become a full member of the group, then it may become a union issue of paid employment. This then raises two issues – the non-payment of a researcher that may have union implications, and the potential impact on an employer to take on an employee.

Unsurprisingly, many employers would be unwilling to take on an unplanned employee, particularly when the employee is likely to be a novice, an adult and hence unlikely to make a substantial contribution in return for her/his salary.

4. DATA COLLECTION

Once access has been negotiated, it becomes necessary to consider the types of data collection to be undertaken. There are three key forms of data collection - participant observation, interviews and collection of artifacts. Millroy (1992) also notes reflection as a method of data collection.

4.1. Participant observation

The key tool for data collection is that of participant observer. Within the hard-core approach, participant observation allows the researcher to become a 'trusted' member of the community and in so doing allows them to experience first hand the social setting from the perspective of the participants (Glesne & Peshkin, 1992). Millroy's study demanded that she became a member of the community of carpenters. As she notes, this took a sustained period of 4 weeks to be accepted within the workshop, with trust and rapport developing over that time. Ball (1988) raises serious issues associated with the notion of what actually constitutes 'participant observation' arguing whether such an ideal is actually possible. While his argument was based on the context of schools and classroom research, he raises issue as to the feasibility of such a position. His general argument is relevant to the workplace contexts as well. Notwithstanding his argument, it is noted that it is an 'ideal' which is strived for in the ethnographic tradition.

Participant-observation demands that the researcher carefully and systematically records aspects of the workplace in fine detail. In the hard-core ethnography, this often demands that the researcher works in the field all day, and then returns to home base to spend the evening writing field notes. However, this approach is not always practical or sustainable within current research environments. In contrast, Ball (1988) proposes that participant observation should be seen to exist in a continuum ranging from observation through to participation. It is likely that throughout the research process, the researcher may assume a variety of locations along the continuum.

Within workplace contexts, the range of positions within this continuum is most likely. Unlike the more common classroom ethnographies, workplaces are constrained by numerous workplace regulations. The contexts of workplaces are substantially different from educational settings. In particular, workplace safety regulations, payment for work, and so on, demand that the employer take responsibility for the researcher. Consequently, it is often impossible to become a full participant-observer within such contexts. In my study of pool builders, there was no possibility of becoming a member of the community due to regulations imposed on the workplace. In this study it was only possible to be an observer with the occasional task of passing equipment.

The participants may also question the status of the researcher. Common folklore within workplaces is sceptical of academics. My experiences with studying numerous workplaces have indicated that most employees are unwilling to share information - particularly when it is related to mathematics – 'the worst subject of all'. This resistance to school mathematics may influence the data collected. In many instances employees are not articulate about their work or processes they undertake, so the presence of a person observing or asking questions can be intimidating. In studying pool builders, I found that the excavators could not articulate how they estimated the amount of soil to be taken from a site. Observations indicated that they were highly accurate with their task, but questioning them only produced 'We just know' and then numerous stories of clients who had no idea of how much soil would be excavated from a site. From the perspective of experts and novices, such stories were the source of amusement for many of the workers, but also served to induct new employees into the discourses of pool excavation. But from the need to gain access to how the participants perceived and understood their work, their resistance (whether intentional or unintentional) to the research, restricted the information which could be gleaned through participant observation. Their resistance or inability to articulate how they estimated the amount of soil to be removed effectively prevented an understanding - from their perspective - of how the task was undertaken. In contrast, my study of bakers (in a supermarket) indicated that they did not see any mathematics in their work, but felt that as my interest was in mathematics, that they needed to talk mathematics with me. Their responses would often include comments about how they did not do any of 'that measuring stuff as it was too complicated for the job they did'. Alternatively, there were comments offered which attempted to integrate school mathematics into their practices, but this was often done as an afterthought. Furthermore, the process of negotiation with the company also meant that supervisors knew the purpose of the study so would pass through the bakery offering comments about the (lack of) bakers' (school) mathematical expertise.

Gender also is an issue for the study of many workplace settings. Many workplaces are very masculinist and the presence of a woman is treated with scepticism, and in some cases, open sexism. Construction sites, for example, are often riddled with entrenched sexism. Sometimes, it may be quite overt, but increasingly with equity reforms, it may be of a more covert form. The latter is more likely to occur in larger organisations which have discrimination policies enacted. However, in smaller organisations where the staff may be less amenable to such reforms, the gender issue is more likely to be apparent. In my work with small groups where there was a heavy emphasis on laboured work, such as the pool builders where there were small teams of men, there were varying degrees of sexism – largely dependant on the individuals. The excavation crew with its heavy equipment tended to have a very entrenched view of masculinity so that it would be very difficult for a woman to take on the role of participant observer. Similarly, there are documented cases of the difficulty men experience when working in early childhood centres, thus making it somewhat more difficult for a man to be a participant observer than a woman in such settings. Such environments make it very difficult to become a 'member' of the group. In these scenarios, the idea of participant observation is far from possible.

The information collected through observations is then compiled into field notes that are often written up at the end of the day. These consist of both observations, but also can include reflections, interpretations and evaluations of those observations. However, it is important that the 'pure' data be clearly distinguished from the other.

The data collected in an ethnography may consist solely from this means, but in most cases it is supplemented with other forms of data gleaned from other methods - interviews or artifacts.

4.2. *Interviews*

Interview questions arise from participation (and/or observation) and are connected to what has become known through that participation. Observing certain behaviours or artifacts allows the researcher to gain a feel for what might be happening. Asking questions about the incident or item allows the researcher to gain access to more information, or to validate what had been noted. This allows for more accurate interpretation of what has been observed and for confirmation from the participant's perspective.

Interviews are most often preceded by observations that provide the impetus for the questions to be asked. In some instances, observations provide fortuitous catalysts for the interview. For example in studying cabinet makers, I noted that the cabinet maker relies heavily on measurements when designing the kitchen, ordering materials and the sizes of the materials to be cut. Further discussions with the cabinet maker confirmed this observation so it was assumed that this was a profession that relied on formal measuring techniques. However, when the cabinet maker went to put handles on the doors, a small wooden device had been constructed which allowed him to drill the holes accurately without making any measurements at all. This observation provided the catalyst for a lengthy discussion on the short cuts used within the profession. The cabinet maker revelled in being able to 'cheat' and not undertake formal measurements. The planning of the kitchen did require formal measurements, but not of the degree initially anticipated. The cabinet maker revealed that there were two levels of operation – the first was that which should be done formally and required quite involved calculations, the second relied on his professional judgment and experience. It was the latter than he had more confidence in. He had many 'tricks' to cover mistakes but also revealed that using and relying on formal measurements was fraught with problems as it made many assumptions about houses. In particular, such calculations assumed that the existing kitchen would be square (that is, all walls would be at 90° to walls and so forth) whereas, in reality, this was rarely the case. This meant that accurate measurements were more often than not, a waste of time and material. It was far more effective to work out 'best fits' and best ways to cut materials (to avoid wastage, reduce costs and hence improvement profit and/or customer satisfaction). If there were gaps, there was a range of strategies that were very effective in masking them.

Interviews can consist of a variety of formats. In the first, and perhaps more commonly used, format, the interviews are very informal and often arise immediately out of the situation. They often occur incidentally - on the job, tea breaks, and so forth - and hence are frequently written in the field notes rather than recorded. Millroy (1992) refers to such interviews as 'informal conversations' (p.86). More formal interviews might be conducted throughout the research process and are more structured. Such interviews can be semi-structured or formal. In most cases these are likely to be conducted on a one-on-one basis, although in some cases it might be useful to conduct group interviews.

4.3. *Artifacts*

Glesne and Peshkin (1992) argue that artifacts provide historical and contextual dimensions to observations and interviews. Millroy (1992 citing Goetz and Le Compte 1984) states that

> anything that the community makes and uses results in artifacts that constitute data indicating people's experiences and knowledge. This may include physical objects like carpentry tools or everyday household objects. (p. 88-89)

The artifacts provide further documentary evidence of the workplace but also can provide the catalyst for interviews. In many professions unique artifacts have been developed from the context. In my study of pool builders, the tradespeople used a variety of tools unique to the context that had been developed from the unique needs of that context. For example, the box-and-framer whose task it is to build the reinforcing frame for the pool, must ensure that the pool is level so that when water is put into the pool the lip of the pool is the same distance around the pool. Contextual cues from the environment are most frequently distracting and inaccurate. A theodolite would be the most appropriate formal tool, but is not used within the profession due to its prohibitive cost and the lay of the land when constructing the frame. The more appropriate tool was a bucket of water with a long length of clear pipe that provided a very accurate measure for the external walls due to similar levels in the water. The level of the water indicated similar heights in corresponding walls. Similarly, the use of formal measuring instruments for measuring certain aspects of the pool is impractical. The box-and-framer uses more 'handy' methods. Building regulations dictate that the lip of the pool must be 9 inches and that the bars in the pool structure must be 12 inches. To measure and check these measurement, the box-and-framer estimated the distance and in most cases was highly accurate. To check his measurements, he used his handspan for the lip (9 inches) and his foot enclosed in his workboot (12 inches). Inches are cited here as these were the units of measure for this aspect of the task although other aspects were in centimetres. This reflects further the use of certain systems of measurement (artifacts) have more value depending on the context of use.

In some instances the collection of artifacts is not possible for any number of reasons. Most people are not too keen about donating their equipment for research! To combat this, I use photographs extensively. I have found that photographs of artifacts, along with the actions being undertaken, provide an excellent stimulus for interviews. In many cases, employees are not keen to discuss what is happening as they are working, but over tea breaks the photographs can be discussed in a very non-threatening manner. As noted previously, one of the major constraints on this form of research is the identification of key informants who are able to articulate their thinking. Where informants are unable to articulate their perceptions or practices, I have found that photographs can provide an excellent stimulus for discussion, often allowing participants to discuss aspects of their workplace which they would not have possibly done.

5. CONCLUSION

Using the methods and methodology of ethnography, permits the documentation of the approaches, skills, and knowledge used within particular contexts in order to understand the reasoning and problem solving skills employed by participants. Within workplace settings, this is a particularly powerful methodology for deconstructing the ways in which participants understand and undertake their daily work practices. This is often in contrast to the methods and understandings developed within formal school mathematics. As the studies cited in this chapter by numerous researchers have shown, participants are often far more effective in undertaking quite complex tasks *in situ* but do not use the formal mathematics learned within the school setting. Ethnography permits a development of understanding of the processes and practices being developed within the particular workplace from the perspective of the participant/s. In so doing, it offers a legitimacy to the non-school-mathematics practices.

Ethnography allows a researcher to develop a theory and representation of the culture and development of mathematical strategies used within that context. In so doing, it makes possible an understanding of the dialectic of context and mathematics. The research emerging from such studies has indicated that different forms of mathematics are developed and used very successfully within workplaces. By understanding the ways in which context and mathematics are mutually constitutive, the definition of mathematics being constrained by school-mathematics is challenged. Such challenges suggest a far more comprehensive conceptualisation of mathematics needs to be developed

The importance of this work is that it challenges current orthodoxies supporting the transferability of knowledge from one context to another. In so doing, it challenges the myth of transposable knowledge and skills, and hence the hegemony of mathematics within the formal school curriculum. Previously, it was assumed that mathematics could be transported from one context to another and as a consequence was a powerful tool. To this end, mathematics assumed a position of high status within the school curriculum and the wider society. In contrast, this emerging perspective suggests that this may not be the case, and that the power of (school) mathematics may be under threat.

The implications of the findings of this form of research are that the mathematics offered in formal school settings may be conceptualised within a very narrow framework and akin to Bishop's (1988) 'Mathematics'. This growing body of research suggests that a much broader conceptualisation of mathematics may be called for due to the various forms of mathematics which develop *in situ* and are not part of the formal school curriculum. For many of the students for whom school mathematics has been a disempowering experience, a broader conceptualisation of mathematics may offer different experiences than the current regimes of school mathematics that are on offer in contemporary institutions. This may be particularly the case for those students undertaking vocationally orientated courses where it is common for students to have low levels of numeracy and have been alienated from the formal schooling processes. The outcomes of this type of research may offer new insights into the forms and pedagogies used in the future teaching and learning of mathematics. Such changes may have a direct impact on vocationally-orientated courses, but may impact on the teaching and learning of school mathematics as well.

6. REFERENCES

Ball, S. (1988). Participant Observation with Pupils. In R. Burgess (Ed.), *Strategies for Educational Research: Qualitative Methods* (pp. 23-54). London: The Falmer Press.

Barton, B. (1997). Anthropological Perspectives on Mathematics and Mathematics Education. In A. J. Bishop, K. Clements, C. Keitel, J. Kilpatrick and C. Laborde, C. (Ed.), *International Handbook of Mathematics Education*, (Vol. 2, pp. 1035-1054). Dordrecht: Kluwer.

Bishop, A. J. (1988). *Mathematical Enculturation: A Cultural Perspective on Mathematics Education*. Dordrecht: Kluwer.

Cooke, M. (1990). *Seeing Yolgnu: Seeing Mathematics*. Northern Territory, Australia: Batchelor College.

D'Ambrosio, U. (1985). Ethnomathematics and its Place in the History and Pedagogy of Mathematics. *For the Learning of Mathematics, 5*(1), 44-48.

D'Ambrosio, U. (1987). Ethnomathematics, What It Might Be. *International Study Group on Ethnomathematics Newsletter, 3*(1).

Denzin, N. (1978). *The Research Act: A Theoretical Introduction to Sociological Methods*. New York: McGraw-Hill.

Eisenhart, M. A. (1988). The Ethnographic Research Tradition and Mathematics Research. *Journal for Research in Mathematics Education, 19*(2), 99-114.

Ellen, R. F. (1984). *Ethnographic Research: A Guide to General Conduct*. New York: Academic Press.

Goetz, J. P., and Le Compte, M. D. (1984). *Ethnography and Qualitative Design in Educational Research*. New York: Academic Press.

Lave, J., and Wenger, E. (1991). *Situated Learning: Legitimate Peripheral Participation*. Cambridge: Cambridge University Press.

Lave, J., Murtaugh., and de la Rocha, O. (1984). The Dialectic of Arithmetic in Grocery Shopping. In B. Rogoff and J. Lave, (Eds.), *Everyday Cognition: Its Development in Social Context* (pp. 67-94). Cambridge, MA: Harvard University Press.

Lokan, J., Ford, P., and Greenwood, L. (1996/7a). *Maths and Science on the Line: Australian Middle Primary Students' Performance in the Third International Mathematics and Science Study (TIMMS) Australian Monograph*. Melbourne: ACER.

Lokan, J., Ford, P., and Greenwood, L. (1996/7b). *Maths and Science on the Line: Australian Junior Secondary Students' Performance in the Third International Mathematics and Science Study (TIMMS) Australian Monograph*. Melbourne: ACER.

Masingila, J. (1993). Learning from Mathematics Practice in Out-of-school Contexts. *For the Learning of Mathematics, 13*(2), 18-22.

Millroy, W. L. (1992). An Ethnographic Study of the Mathematical Ideas of a Group of Carpenters. *Journal for Research in Mathematics Education, Monograph 5*.

Mitchelmore, M. (1996). Introduction. In J. Mulligan and M. Mitchelmore (Eds.), *Children Number Learns* (pp. 1-16). Adelaide: Australian Association of Mathematics Teachers.

Nunes, T., Schliemann, A., and Carraher, D. (1992). Ethnomathematics and Everyday Cognition. In D. A. Grouws (Ed.), *Handbook of Research on Mathematics Teaching and Learning* (pp. 557-574). New York: Macmillan.

Rist, R. C. (1980). Blitzkreig Ethnography: On the Transformation of a Method into a Movement. *Educational Researcher, 9*(2), 8-10.

Robitaille, D., and Nicol, C. (1994). Comparative International Research in Mathematics Education. In R. Beichler, R.W. Scholz, R. Straesser and B. Winkelmann (Eds.), *Didactics of Mathematics as a Scientific Discipline* (pp. 403-414). Dordrecht: Kluwer.

Saxe, G. (1988). Candy Selling and Mathematics Learning. *Educational Researcher, 17*(6), 14-21.

Spindler, G., and Spindler, L. (1987). Teaching and Learning how to do the Ethnography of Education. In G. Spindler and L. Spindler, (Eds.), *Interpretative Ethnography at Home and Abroad* (pp. 17-36). Hillsdale, NJ: Lawrence Erlbaum Associates.

von Glasersfeld, E. (1984). An Introduction to Radical Constructivism. In P. Watzlawick (Ed.), *The Invented Reality* (pp. 17-40). New York: Norton.
Watson, H. (1988). Language and Mathematics Education for Aboriginal-Australian Children. *Language and Education, 2*(4), 255-273.
Wolcott, H. (1987). On Ethnographic Intent. In G. Spindler and L. Spindler, (Eds.), *Interpretative Ethnography at Home and Abroad* (pp. 37-57). Hillsdale NJ: Lawrence Erlbaum Associates.
Zevenbergen, R. (1996). *The Situated Numeracy of Pool Builders*. Critical Forum.
Zevenbergen, R. (1998). Ethnography in the Classroom. In J. Malone, B. Atweh and J. Northfield (Eds.), *Research and Supervision in Mathematics and Science Education* (pp. 19-39). Mahwah, NJ: Lawrence Erlbaum Associates.

7. NOTES

[1] While there is a politically-driven shift back towards wide-scale testing and performance, this is actively resisted by many researchers concerned more with developing theoretical and/or cognitive models of mathematical understanding from the perspective of the participant – whether the young child, adolescent or the worker.

[2] See Millroy 1992 for a succinct precise of this debate.

CHAPITRE 16

VISIBILITY OF MATHEMATICAL OBJECTS PRESENT IN PROFESSIONAL PRACTICE[1]

ANNIE BESSOT[2]

Abstract. On a building site, the operations carried out by the workman are the most visible part of his work. These actions, which have a professional purpose, are the outcome of a body of knowledge which is less visible to a building site observer (because it is unformulated and is often incapable of being expressed). This knowledge forms part of the device which makes it possible for the actions observed to take place. We describe a methodology which allows mathematical objects present in professional pratices to be made visible. An analysis *a priori* of possible techniques for creating the formwork of a wall on a sloping slab fed into an investigation on a construction site course which attests to the presence of mathematical knowledge within the structure of vocational training. The problem of assuming responsibility for this knowledge within vocational education forms part of the discussion of this paper.

In the context of this article, we are concerned with the presence and nature of a working knowledge of mathematics[3] in representative professional techniques: we have chosen professional techniques in the building industry which are linked to *actual building tasks*[4].

1. ACTIONS AND DEVICE, A DIVISION BETWEEN THE VISIBLE AND THE INVISIBLE

Chevallard (1991; see also Mercier in this section) suggests creating a division within a practice between actions and device:

> [...] when we divide up the activity within a practice, there is a large-scale withholding of certain elements which determine whether this practice can be carried out. [..] It is the actions of those involved in the practice which can be seen most clearly on a cultural level [...] There will on the other hand be a strong inclination to forget the other component in a practice, which is less visible but just as essential, and which is what I call the device. The device includes the whole set of objects and processes which allow the practice to be carried out. (op. cit. pp. 105-106)

In the professional area being studied, the part of the device made up of technical tools determines professional practice in the same way as do the succession of actions. Mathematical knowledge is preserved in these tools, and adapted to specific situations. But is it not also an *invisible* tool? One of the concerns of vocational

Bessot & Ridgway (eds.), Education for Mathematics in the Workplace, 225—238.

training is to determine what mathematical knowledge is workable in practice. In order to be presented as subject-matter for teaching, it must be made visible and be capable of being expressed: this is one of the conditions for deciding whether or not mathematics can be taught. Are there particular points at which mathematical knowledge is formulated, thus making it visible for a time as a necessary precondition for tasks to be carried out on the construction site?

In order to try and identify such moments, and in accordance with our line of questioning, we have carried out our research in a place which combined the characteristics of being a construction site, and thus an ideal observation-point for studying practices used, and at the same time a training ground, where the information that the pupil needs to know is set out and formulated.

The order was for a delivery of ten garages for a housing development being constructed by a building firm at the same time. As with any construction site of a certain scale, consideration for the project's layout was referred to the judgement of the architect during the course of carrying out the project. In addition to their training purposes, the construction site courses were part of the assessment procedure for obtaining the vocational diploma of building site supervisor[5].

In this construction site training course, as in any training situation, for both the teacher and the pupil, it is not enough to know how to do a task, it is also necessary to be able to justify and explain what is being done: it therefore follows that the setting up of a training plan facilitates both the formulation of learning objectives within certain prescribed areas and, closely linked with this, the smooth development of the practical procedures involved.

Where methodology is concerned, our research was guided on the one hand by a theoretical analysis designed to identify the points at which such objectives are formulated (*a priori* analysis), and backed up on the other by a system of gathering observation data.

2. AN *A PRIORI* ANALYSIS OF SITUATIONS IN WHICH A WALL IS CONSTRUCTED

A priori analysis is both a methodological approach and a result.

> [It] does not refer back to a position in time relating to the history of research but to thinking which is of an epistemological nature, taking into account the general situation and not the specific situation which will be produced. [...] Of course, a priori analysis does not allow everything that may occur to be foreseen: it allows the data for observation to be separated out according to the questions set by the research. (Bessot and Comiti, 1985)

2.1. *Need for checking*

The effectiveness of construction site practices is dictated by a plan of contract between client, architect and those who occupy different positions in the building firm.

On modern construction sites (like the one where the construction site training course takes place), prefabrication and minimising of costs impose new constraints: components prefabricated outside the construction site must be adapted to fit in with work already carried out on site. These conditions mean that a construction site must include in its practices different checks on the appropriateness of measurements.

Certain checks are in the nature of practical problems:

> The action whose results are immediately visible and continuously assessed is the principal means of control for solutions to corresponding situations: if the solution is not satisfactory, it will be modified by adjustment until it becomes so. (Berthelot and Salin, 1992)

Others are connected to the problem of creating models[6] which is linked to Euclidean geometry:

> [...] it is clear [...] that all of these corrections are made *according to the Euclidian model.* Errors and corrections are determined with the - usually unconscious - intention of rendering the whole system of measurements capable of interpretation through Euclidian geometry with a constantly narrowing margin of discrepancy between the figures. It is very difficult to imagine how one should proceed if this abstract model [...] is not to be the guide for all the measurements which are to be carried out. (Gonseth, 1926)

With these problems in mind, checks may be made at two stages:

-work carried out may be checked: checking is done to match the purpose of the action against the current state of the work. Checking possibly entails a decision to make an adjustment. We shall henceforth call this *verification*;

-a check may be made on the validity of an action to be carried out by *anticipating* the actions, difficulties and errors which might arise: this form of checking usually requires calculations of measurements within the tolerance limits in order to guarantee the quality of the work and minimise costs. It is a check for which mathematical knowledge provides the tools.

It is at these two levels that we propose to try and identify points at which mathematical knowledge becomes visible (for an another methodology, see Noss and Hoyles 1996).

2.2. Casing for a wall perpendicular to an existing wall

Once the markings showing the position of the wall have been drawn, formwork moulds to build the wall are installed. If the ground surface is '*plane*' and '*horizontal*' this is a routine construction site task. The moulds are mainly made up of two parallel wall forms: one wall form is made from a metal plate mounted on a frame to ensure that the plate remains rigid. The dimensions of a wall form are standard: generally, the construction of a wall requires several successive wall forms - *the building length is then a multiple of the length of a wall form.* The pairs of wall forms which constitute the mould are held in place and adjusted vertically by mechanical means. As they are very heavy, these wall forms are moved by tower crane. This means that the smooth running of operations must be checked by the setting-out lines. As an aid to understanding for readers unfamiliar with construction site procedures, the position of the wall forms in the situation described is illustrated in Figure 1.

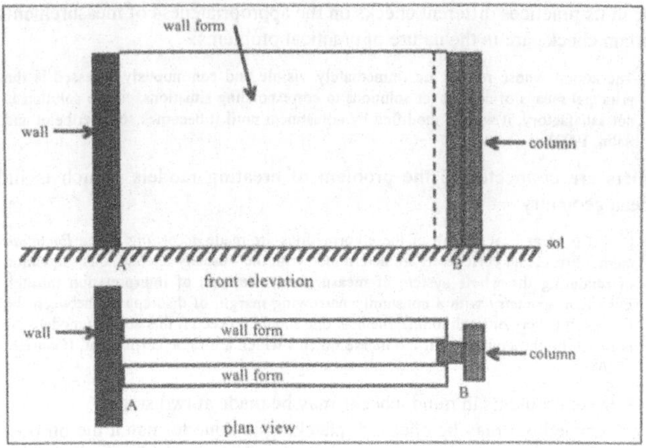

Figure 1. The position of the wall forms

2.3. *When formwork becomes problematic*

A second situation arises when the 'plane' surface of the ground *slopes* to allow water to drain away: in this case, the slope may be indicated by a percentage either in the detailed estimate of the work or on the construction plan by an arrow as follows:

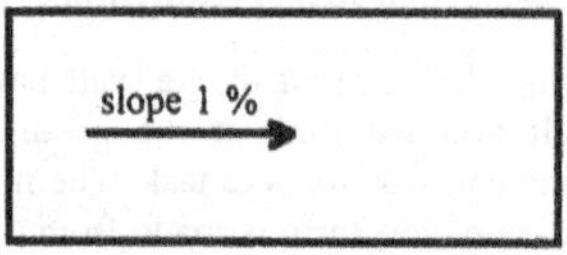

In this way the point and the place at which the construction task becomes problematic can be spotted in advance, and the risk of error will require checks to be carried out more frequently at both levels. In the construction site course, one of these points is when a 7 metre-long wall is built on a slab with a *slope of 1% with the horizontal plane* (Figure 2).

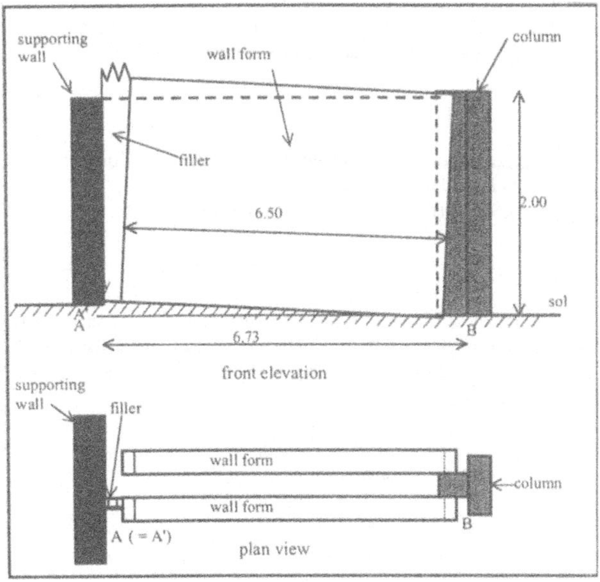

Figure 2. The position of the wall forms on a slab with a slope of 1% with the horizontal plan

On the opposite side to the supporting wall, a prefabricated column is put in place against the wall forms. Its shape allows it to be housed into the wall forms in such a way as to close the mould (on the column side). The position of this column provides a marker (B) for laying the wall forms.

But a space, which may be large or small in size, remains to be filled between the supporting wall and the wall forms: *it is here that the problem is located!* It is therefore necessary to construct a tailor-made mould, called a filler, so as to extend the wall forms. This filler is made up of a plate in plywood ('sheeting' of the formwork) with a wooden frame. Its purpose is to ensure the continuation of the wall forms.

2.4. Possible techniques for constructing the filler

Three are set out: the second is not found on the construction site and the two others are to be found under different conditions.

2.4.1. Two techniques using calculations

These techniques (1 and 2) require prior calculation of the filler's measurements. This calculation can be made outside of construction site working time, from the plans. Either the architects'office[7] or the site foreman takes responsibility for this.

The y calculation (see Figure 2) *is common to both techniques:*

$$A'B = y + 6.50 + x$$

On the plan, measurements are always given on horizontal or vertical directions. The dimension AB is read on the plan.

The approximation[8] AB H A'B gives a measurement of A'B

$$A'B \approx 6.73\ m$$

The calculation of x is based on the correspondence[98] between the slope of the slab and the false plumb of the wall form. The slope (read on the plans) is 1%.

The 1% slope means 1 cm for 1 m. So for the false plumb one has 2 cm for 2 m, from which x = 2 cm is arrived at.

$$x = 0.02\ m$$
$$A'B = y + 6.50 + 0.02$$
$$y \approx 0.21\ m$$

Once the calculation is made, how is a plywood plate cut out in order to construct the filler?

2.4.1..1 Technique 1, a technique which controls the transition from model to actual object

After choosing a suitably-sized plate (that is one which is longer than 2 m and well over 21 cm in width), the actions necessary for marking out (before cutting out) consist in transferring measurements of length: 21 cm; 2 cm for 2 m (or 1 cm for 1 m) as in Figure 3. The cutting out is done with the help of a mechanical saw as used in a workshop[10], which is not available on the construction site.

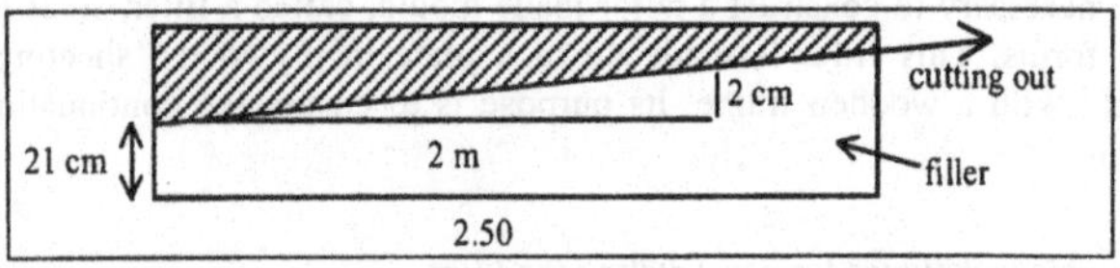

Figure 3. The marking lines before cutting out (technique 1)

Technique 1 *anticipates that the filler will require rectification* at the point when it is constructed on site following verification. This expectation is evidence of the fact that in this technique, the relationship between the Euclidian model within which the calculation has been carried out and the reality within which the action will be carried out has been taken into account.

The rectification rests on the use of a construction site mechanical saw which comprises a guide allowing cutting out to be done according to a parallel plane *on one edge* of the plate only. We set out below a diagram of two rectifications according to whether y is too wide or too small.

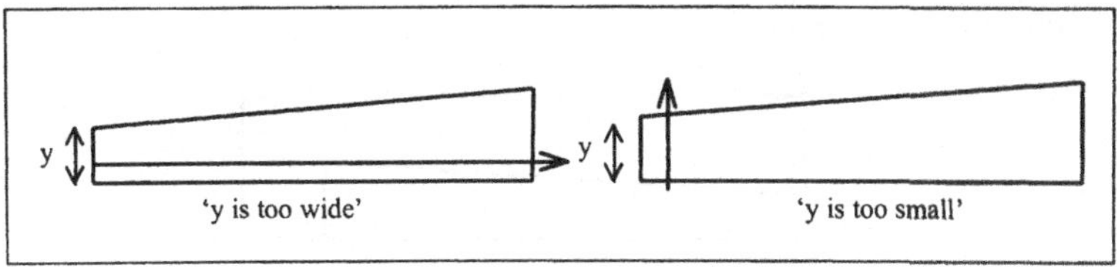

Figure 4. Cutting out in order to rectify (technique 1)

2.4.1..2 Technique 2, an imaginary mathematical technique

The actions for marking out on a plywood plate (before cutting out) consist in transferring measurements of length 2 cm for 2 m (or 1 cm for 1 m) as on Figure 5. A right-angled triangle is obtained of which one acute angle α is such that $tg\alpha = 1/100$. h is calculated to obtain a right-angled triangle such that the opposite side to α is of y length equal to 21 cm. Then the point of the triangle is eliminated to obtain a trapezium whose small base has y as its length: this trapezium represents the desired filler.

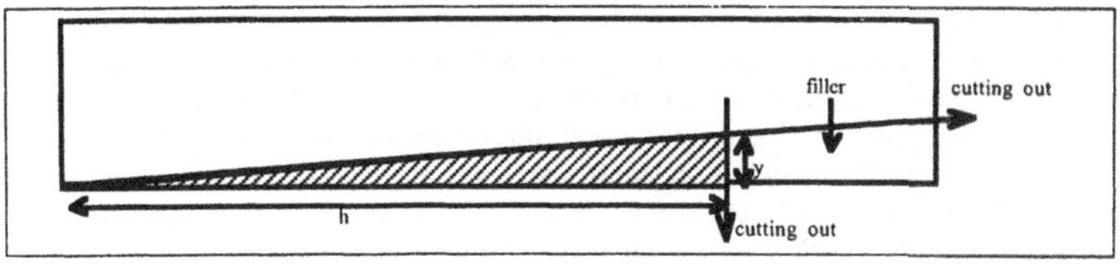

Figure 5. The marking lines before cutting out (technique 2)

This technique is the one which we believed to be in use on the construction site under observation. In fact, it is never used, and this is evidence of the impenetrability of practices divorced from the mathematical traditions of institutions with which we are familiar. This problem was resolved in the same way as an elementary mathematical problem *without making an actual link* with the reality represented by a model: it was enough for us to formulate the calculation in order to be assured that *it was possible to build the filler*. If h is calculated effectively, h is found to equal 100y, hence h = 21 m, or 21 metres of waste! The mere cost of materials makes this technique non-viable, and where would such a plate be found?

2.4.2. A technique 'without calculation': technique 3, a pragmatic technique
On the construction site, once the wall form is in place, the filler can be built *without using the slope* given on the plan. One proceeds by 'direct measurements' L1, L2, L3 and by transferring these measurements onto the plywood board (L' is a measurement resulting from this).

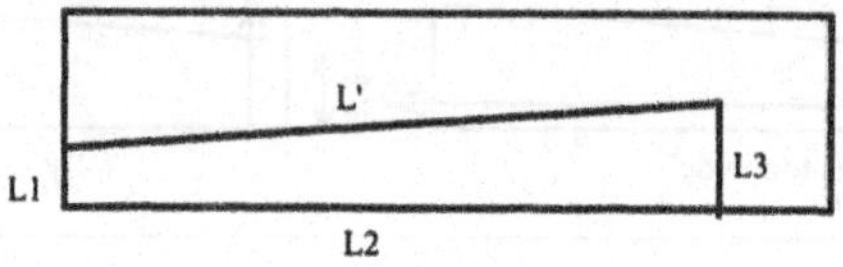

Figure 6. The marking lines before cutting out (technique 3)

2.5. Operating conditions for techniques found on the construction site

Technique 3 is a pragmatic solution in which no part is played by a model which can be calculated. It can provide a solution when conditions making it desirable to use technique 1 are absent. This adjustment on the construction site, if such proves necessary, may be enough for practical purposes, for example in the case of small-scale construction of a single wall.

But the quality standards for cutting out and for putting the framework in place require specific equipment and materials, and demand that the cutting be done outside the rushed conditions of the construction site itself, using a 'cutting workshop' with precision tools for marking out lines and cutting (for example, the workshop saw). On the construction site, the tools available do not have the same function: tasks like cutting the *oblique L'* out of the board are rendered extremely delicate because the *guide for the construction site mechanical saw* cannot be used. What is more, measurements taken on the construction site are most unreliable when constructing a very high wall or when the ground slopes steeply.

Technique 1 rests directly or indirectly upon the correspondence between the slope of the slab and the false plumb of the wall form. A necessary condition for technique 1 to work is that the construction firm has sufficient *knowledge* to foresee the measurements of the filler from a reading of the plans (calculation of x and y in our analysis). In the absence of this knowledge (which relates to the Euclidian model and to the notion of slope), the requisite demands for quality will be difficult to satisfy through a pragmatic technique which can only operate once the wall forms are installed.

3. INVESTIGATION OF A CONSTRUCTION SITE TRAINING COURSE

What emerged from the preceding *a priori* analysis fed into our investigation and the interpretation of the data gathered. The way in which data-gathering for this experiment was organised is described below.

3.1. Organisation of the investigation into the construction site training course

The researchers followed one or other of the trainers on the construction site and questioned them as if they had no knowledge of what was going on. Questions and answers were recorded[11].

This organisation results from the decision to look at the construction site through the questions and answers of people holding positions within different institutions: on the one hand researchers, who were unaware of construction site practices, on the other construction site professionals and teachers from a technical high school. Because of the lack of knowledge on the part of the researchers, this variety of standpoints allows actions which are taken for granted, and are thus transparent for professionals, to be questioned. The *professional* questioned in this way is led to justify practices and make them understandable, and thus to formulate knowledge which can be used in these practices. The *teacher* indicates for these same practices what the pupils must be capable of doing in relation to the training plan.

On the construction site, the pupils' work[12] was supervised by two people who we will call R and G.

3.1.1. The two professional positions of R and G

As professionals R and G have different specialisms: R has worked in the building industry for a year as a foreman, and as site manager[13] for five years. He is responsible for the construction site training courses. G was an independent technical adviser to architects and building firms; he is concerned more with setting out and with the overall organisation and time management of tasks[14]. G expresses his lack of knowledge of the formwork procedures that we have just been observing:

> G: It's a job I've never done, so I'm not very comfortable with it. There's only my colleague who can do the adjustment. Anything to do with setting out I know how to do; but formwork, I've always let the people on the worksite take care of that. [...] *I'm learning as much as they are* (the pupils).

We shall examine later whether this last statement is valid.

3.1.2. The two teaching positions of R and G

R and G are both teachers of construction but R is responsible for what on the timetable is called 'workshop'[15], and the construction site training course is part of this. If G plays a part in the workshop, his teaching relates more to knowledge appertaining to the architects' office.

Once work on the construction site was finished, the investigation was completed by a questionnaire concerning the progress of work on the site, designed for the pupils as well as the teachers, R and G.

3.2. An episode from the construction site training course or a forgotten rectification

A brief episode[16] was related concerning the installation of the filler.

It was in the high school, during the period when pupils were being taught, that the problem of constructing the filler arose: the teacher R took responsibility for calculating y by showing the actions involved in this calculation to the pupils. The pupils' work was confined to marking out the lines and cutting out the filler on the basis of the information y = 21 cm. G (a non-specialist) attributed the error to the calculation, whereas this unavoidable error was a result of the transition from the geometrical model to the reality of the construction site.

> G: The part that had been earmarked in the workshop to take up the vertical slack was wrongly calculated, *there are 2 centimetres to take off*

It is the case in the following situation, foreseen in technique 1:

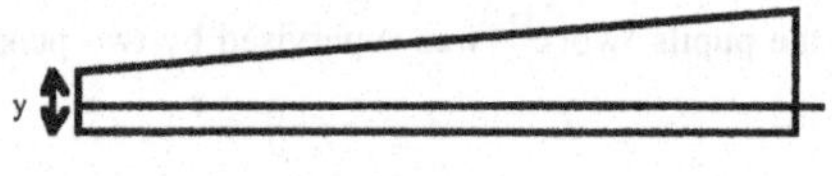

> A[17]: That's interesting. [...] Who's going to correct it, the pupils or the teacher?
>
> G: The teacher marks out the lines and is going to cut the piece of wood with the saw from the building site. The error appeared when the mould was laid on the slab [*when verification was taking place*], and it didn't correspond to the line markings because the end went beyond its normal position by 2 centimetres.
>
> A: And will the mistake be rectified by making fresh calculations?
>
> G: They've marked out the lines. The teacher corrected his error on the spot by taking the real measurement [*by direct measurement*]
>
> A: [...] and not by calculating?

Since at the time of observation A only imagined a 'mathematical' technique being involved (technique 2), she was incapable of envisaging any other rectification device.

> G: No. In fact the calculation was right, for the angle, the slope, but not for the length. There is a problem with the length [...]

The cause of the error is not spelt out by R because for him it really is an unavoidable error; he is able to list the various basic actions which are mandatory if precision is to be achieved (prior marking out, for the cutting and the laying of the wall forms). Responsibility for rectifying the error which is an integral part of technique 1 is taken on entirely by R. This will have consequences for the relationship which the pupils and G have *vis-à-vis* the knowledge involved in technique 1.

The 2 cm error was noted by everyone. After the event, the error would be forgotten by R and the pupils. We asked the following question:

> Did you encounter problems when setting out the wall forms on the sloping slab for the walls separating the garages? Yes no / If so, what were these problems?

This question received a negative response from R and all the pupils. Only G remembered! The memory lapse of R is of course of a different nature to that of the pupils. His forgetfulness is that of a professional for whom rectification and the knowledge justifying it are taken for granted; for the pupils, the error was not perceived and the actions necessary for rectifying it have not been learnt. As for G, he remembers the 'rectification' of the unavoidable error because he has learnt it.

The anticipated construction of the filler formed the subject matter for teaching, over and above the time spent on the construction site, but the pupil's work was reduced to the actions necessary for carrying out the technique: calculations and anticipation of errors which are unavoidable and acceptable within the margins of tolerance remained the job of the teacher, whereas these calculations and anticipation of error are part of learning how to be professional. As a teacher of construction, R considers that he does not have responsibility for teaching the mathematical knowledge which would allow construction and rectification of the filler to be carried out: he has this knowledge but considers it to be elementary. The teachers R and G refer to this knowledge by the term 'understanding of slopes'. Conversely, without this knowledge there is no assurance that technique 1 can operate, and the possible solution of technique 3 is the only option open to professionals.

Let us clarify what the idea of a slope means within the framework of trades in the building industry before examining where and when the mathematical knowledge is taught which is linked to an understanding of it.

4. THE FORMS TAKEN BY THE NOTION OF A SLOPE WITHIN INSTITUTIONS

4.1. The notion of slope within trades in the building industry

Only the tilt which relates to a horizontal or vertical plane is indicated; the tilt relating to an 'oblique'distinguish the batter, which indicates the tilt in relation to the vertical line, and the slope which indicates the tilt in relation to the horizontal. On architects' plans, the slope is usually *written* as a percentage. This percentage is *read* with the unit system: '1 centimetre for 1 metre' (for 1%). The word 'slope' thus has a dual meaning: tilt on a horizontal, and number characterising this tilt. Let us examine the latter meaning.

4.2. The notion of slope in 'vocational' teaching

For technology teachers, the connection between slope and the mathematical notion of tangent is necessary because it allows access to the angle-tilt. They thus describe a mathematical task which is difficult for Vocational Diploma pupils: 'Let us say that a slope is 47%. Give the value of the angle.' Actually, the link between slope and tangent demands that 47% be interpreted as a number and that this number is transformed when written down. This is a difficult operation for pupils: 47% (read as 47 cm for 1 m) is also 47/100 or 0.47. Information as to the value of the angle can then be obtained by calculator: all that remains to be done is to enter the number 0.47 and press the key which is the inverse of the tangent in order to arrive at the angle.

We may at this point ask whether these relationships between slope, percentage, tangent and angle exist in maths teaching in France.

4.3. The notion of slope in maths teaching in technical high schools for the building industry

In France, the subject of Trigonometry appears in first year maths syllabuses in technical high schools. The tangent of a real number is defined as a function based on the relationship sinx/cosx. It therefore plays a minor role in comparison with the functions of sine and cosine. Nothing in the set texts for the syllabus or in most of the current textbooks proposes the creation of geometrical models which would justify treating the architect's slope as the tangent of an angle. The graphic representation of a linear function - $y = ax + b$ - is a straight line, one of whose characteristics is its angle in relation to the axe of the x-coordinate which is horizontal, that is to say the slope, but also the coefficient a as well as the tangent of the angle made by the intersection of the vertical and horizontal line. This might give rise to work being done in mathematics on the notion of slope. But the study of trigonometric functions is too oriented towards the study of behaviour of functions for such work to be introduced.

This situation of not taking on responsibility for access to the angle-tilt breaks with the situation which prevailed in France prior to the 1970's. In works of that period (for example Oriol, 1964) of an equivalent level, the writings of the same number 24/120 = 0.20 or 20% were based on notions of decimal fractions and ratios of size. The coherence of the relationship between slope and tangent (allowing access to the angle) was reinforced by the introduction of the working drawing, which served as intermediary between the reality of the professional world and the field of mathematics.

5. CONCLUSION

Resolving a problematical task on the construction site requires checks on two levels: anticipating actions in a situation where models are being set up, and verifying the results of actions in the practical area. The example used in this article was the creation of formwork for a wall on a sloping slab. We have shown that mathematical knowledge forms part of a device creating conditions for the optimum technical solution to be developed. In the high school, the actions involved in calculating, marking lines and cutting out (the visible part of the technique) have been taught. But transfer of responsibility for these actions to the pupils was only achieved in the practical sphere (marking lines and cutting out). Because of this an essential part of the professional technique could not be learnt by the pupils since it was not made visible to them in the teaching. The teacher does not consider it his/her responsibility to pass on knowledge which would justify and allow calculations to be made. This knowledge, pertaining to the relationship between slope, tangent and angle within a Euclidian model of space, disappeared from mathematics in the 'professional' branches of the building industry in France in the 1970's. Since then it has never again been officially taught in any other context.

A question opened up by its disappearance is whether it is legitimate in maths teaching to transmit knowledge which has been transformed by professional problems.

6. REFERENCES

Berthelot, R., and Salin, M.-H. (1992). L'enseignement de l'Espace et de la Géométrie dans l'Enseignement Obligatoire. University of Bordeaux I.

Bessot, A., and Comiti, C. (1985). Un Élargissement du Champ de Fonctionnement de la Numération: Étude Didactique du Processus. Recherche en Didactique des Mathématiques, 6(2.3), 305-346.

Bessot, A., and Eberhard, M. (1995). Le Problème de la Pertinence des Savoirs Mathématiques pour la Formation aux Métiers du Bâtiment. In G. Arsac, J. Grea, D. Grenier and A. Tiberghien (Eds.), Différents Types de Savoir et leur Articulation. Grenoble: La Pensée Sauvage.

Chevallard, Y. (1991). Dimension Instrumentale, Dimension Sémiotique de l'Activité Mathématique. Paper presented at the Seminaire de Didactique des Mathematiques et de l'Informatique 1991-1992, LSD2-IMAG, Laboratory, Grenoble.

Gonseth, F. (1926 (new edition 1974)). Les Fondements des Mathématiques. De la Géométrie d'Euclide à la Relativité Générale et à l'Intuitionnisme. Paris: Blanchard.

Margolinas, C. (1993). De l'Importance du Vrai et du Faux dans la Classe de Mathématiques. Grenoble: La Pensée Sauvage.

Noss, R., and Hoyles, C. (1996b). The Visibility of Meanings: Modelling the Mathematics of Banking. International Journal of Computers for Mathematical Learning, 1(1), 3-31.

Oriol, R. (1964). Mathématiques de l'Apprentissage. Paris: Dunod.

7. NOTES

[1] Certain sections of this article are based on the article by Bessot & Eberhard (1995) and link up with the article by Eberhard which appears in section 1 of this publication.

[2] This article is based on research carried out by the following team: Bessot, Déprez & Eberhard.

[3] For a distinction between working knowledge and basic knowledge see Eberhard in section 1 of this publication.

[4] *just after* the tasks of setting out or marking out (see the articles by Eberhard (section 1) and Bessot (section 3) in this publication).

[5] In France this diploma, which is of the same level as the baccaulauréat, leads to intermediate supervisory posts on the construction site, that is posts of 'team foreman'.

[6] In the creation of models, the validity of solutions is guaranteed by the inclusion of rules which are inherent within the model.

[7] The place where technical problems relating to the building project are solved.

[8] Justification for this approximation is given by G as follows:
'The error committed by this approximation is slight. The proof is if the angle is noted by α (AB,A'B).
$\cos\alpha \sim 1 - \alpha^2/2$; or $\alpha \sim tg\alpha = 1/100$; which results in $\cos\alpha \sim 1 - 1/20000 = 19999/20000$;
$A'B = AB/\cos\alpha \sim ABx(20000/19999)$; $e = AB\text{-}A'B' \sim AB/19999$'.

[9] This spatial relationship can be expressed in terms of equality between angles with orthographic sides on a front elevation like the one we have given in Figure 2.

[10] Circular saw with moving table and guide for angles.

[11] We followed the construction site training course regularly over a period of two months.

[12] The pupils generally carried out orders, and occasionally took on the position of team foremen.

[13] General supervision of the construction site.

[14] That is planning tasks, ordering materials and equipment, organising the workforce and security of the construction site.

[15] In France the term 'workshop' marks a separation, in vocational training, of the part which is called practical from the part which is called theoretical.

[16] The term episode is used in the sense given to it by Margolinas (1993): 'An episode is dependant on a particular observation, and is a reconstruction made entirely after the event. Determining whether something is an episode is a matter for analysis of protocol'. (p. 74).

[17] A indicates Annie Bessot.

CONCLUSION

CONCLUSION

RUDOLF STRAESSER

Obviously, there are different ways to write a conclusion of a book which consists of more than a dozen of papers grouped into four sections on 'mathematical knowledge in school and work', 'bringing school and workplace together', 'teaching future workers' and 'research methods for mathematics at work'. I will not go through all the wealth and richness of each and every paper in the book but will try to step aside and put forward my view on four topics. These topics to me seem most important when looking into workplace related mathematics and its teaching - and I will not refrain from taking up arguments which have been formulated in the papers, even in contrasting them with other or my personal views.

Nevertheless and at the very beginning of the conclusion I want to point to one major shortcoming of research (of a multitude of shortcomings) done in the field: 'workplace related mathematics and its teaching' obviously does not take into account the *subjective, learning* part of workplace mathematics. And - like the other authors of the book - I am not in a position to fill in this gap in our scientific knowledge - even if we all know that learning workplace related mathematics takes place everyday all over the world and is a presupposition of the whole book.

1. MATHS AT THE WORK PLACE

In the present book and in other places, lots of facts and arguments can be found about the role mathematics plays in vocational contexts, at work. Even with this amount of literature one has to point to a major difficulty still not really overcome: How to get a valid, realistic description of the role mathematics plays in these contexts? Even with very different ways to analyse this question (from the traditional interview and analysis of syllabi in the research studies of the Cockcroft-report, the on-site research described in the paper by Hoyles&Noss to sophisticated ethnomethodological approaches), there is no simple answer available. It is rather difficult to extract the mathematics from the workplace contexts. Mathematics seems to hide itself in the workplace and special attention and force is to be invested to find, identify and describe workplace related mathematics. In addition to this, specific evolutions are linked to the growing use of technology which adds to this process of hiding mathematics from societal perception (see also the part on technology further down). As for details, there is evidence from other sources that workplace mathematics is not restricted to arithmetic and elementary geometry.

Bessot & Ridgway (eds.), Education for Mathematics in the Workplace, 241—246.

Stimulating examples of rather demanding mathematics used in the workplace can be given, including numerical analysis and optimisation. And instances of data-gathering, data analysis and presentation which sometimes heavily rely on qualitative, geometrical concepts can be identified.

Most of the work on this topic - at least as presented in this book - also remains within a major restriction: in most cases, the vocational contexts of qualified workers are analysed. Very few information is synthesised and taken into account for the mathematics used by experts with an academic background. The more one climbs up the qualification ladder, the less information is available.

In other places (like for instance an ICMI study on the topic), there is an ample discussion on the teaching of mathematics 'as a service subject' - and a widespread complaint that this teaching is often done in separation from the necessities of the respective profession. Is it more than a speculation that exactly this non-utilitarian way of teaching is a way to cope with the definition of workplace related mathematics? Is it a viable procedure not to bother too much about the actual necessities of the workplace? And how to select an appropriate workplace related mathematics for future qualified workers if the actual workplace does not offer the criterion to decide on the mathematics taught? Can (and must) we also rely on a more formal legitimisation of vocational mathematics by putting the general competency of mathematising into the foreground of vocational mathematics (even at the qualification level of future qualified workers) and not so much the mathematics actually or foreseeable in use at the workplace?

Besides these more or less objective facts, findings and arguments, the paper by Wegede in section 2 of this book is a reminder of an often forgotten aspect of mathematics at the workplace: Even before or beside the teaching and learning of workplace related mathematics, the 'simple' use of mathematics at the workplace is already deeply influenced by the personal, subjective relation of the user to mathematical concepts and procedures in use at her/his workplace - be the mathematics hidden or easily identified. The individual has to cope with the 'relevance paradox' of Niss on the personal level which can be described as the feeling 'Mathematics is useless to me, but at the same time I know that I am useless without mathematics' (cf. Niss 1994, p. 377). There is surely a point in studying the personal, subjective aspect of workplace related mathematics.

2. MATHS FOR WORK TAUGHT IN DIFFERENT SETTINGS

As represented in the papers of this book, there are two major types of institutions to teach workplace related mathematics: the classroom and the workplace itself (and I will not go into comments on the distribution of this two institutions in different nations or continents). The distribution of teaching (and learning) workplace related mathematics between two different institutions with different situational, organisational and social patterns immediately implies a problem studied in mathematics education since its birth as a scientific discipline: how to promote transfer of competencies from one institution to the other - and I deliberately take the symmetric wording because transfer from the (novice's) workplace to the vocational college classroom is as important as the transfer from the mathematics classroom of this college to the workplace of the student. In the book, we can find a more pessimistic perspective of the transfer from classroom to the workplace in the paper by Noss&Hoyles, whereas Evans offers an active approach to willingly and explicitly build bridges between the two settings in order to facilitate the necessary transfer.

At a first glance, the institutions (workplace and classroom) are clearly linked to two pedagogies, namely legitimate peripheral participation for the workplace and a modelling approach for the classroom. I will not describe the approaches and their respective distribution in the reality of the vocational training and education (for this cf. Straesser in section 1 of the book), but I like to invite the reader to think about two (thought) experiments on the application of respective pedagogies: The imitation, if not introduction of a community of practice into classroom type of teaching does not come as a total surprise: the discussion on 'authentic' learning (as can be exemplified by the paper by Forman&Steen in section 2) as well as the efforts to know better about the actual workplace mathematics come from the wish and will to be as near as possible to the workplace and legitimate peripheral participation within the classroom (if there is classroom teaching left at all in vocational training). Recent developments in business education (like the simulation of workplaces within business colleges, the 'Uebungsfirmen' in Germany) come from the very same ideas of communities of practice and legitimate peripheral participation - even for the teaching of workplace related mathematics. Nevertheless, I want to add two remarks which show the limits of this approach already under experimentation: It should be clear that classroom type teaching is different from the actual workplace, has to follow different rules and has to cope with the 'didactical transposition' (cf. Chevallard 1991) which implies rules of the game for institutionalised learning which differ from actual social practice at workplaces. The second remark comes from the analysis of the use of mathematics in the actual practice: Nunes et al. (1993) showed in their analysis of 'street mathematics' that classroom type teaching at least has the potential to overcome the limitations of the actual practice. It can offer a broader perspective on work which may be of help when the (qualified) worker has to cope with unfamiliar, unexpected or even 'breakdown' situations. To put it the other way: There may be good reasons not to narrow down the knowledge taught to the actual practice at the workplace - and this additional knowledge may be best taught not in communities of practice, but in institutions like classrooms separated from the workplace.

My last remark already starts the second thought experiment which looks into the opposite direction: What about introducing classroom type training into workplace settings? The idea is not as strange as it may look at the first glance: In technologically advanced workplaces (e.g. in certain types of mechanical construction, design of electronic circuits and the like) exploration by simulation seems to be the only way to introduce novices into the workplace practice - and even experienced workers heavily rely on techniques like these in order not to endanger and damage expensive material and machinery. If society is interested in highly qualified, self-reliant workers it may be worthwhile to continue training and education beyond the narrow necessities of the actual workplace - and this can best be done in a learning/teaching setting which is not too narrowly bound to the actual workplace, maybe even of classroom type.

3. ROLE OF ARTEFACTS / TECHNOLOGY

For mathematics at the workplace and in teaching and learning workplace related mathematics, various papers in the book show the importance not only of mathematical concepts and procedures, but also and sometimes as important as the mathematics the role of artefacts for workplace mathematics. These artefacts can be man-made material devices like tools or machinery, but also ways to organise the (re)production and distribution of goods, even human competencies can be seen as (primary) artefacts (for a broad definition of artefacts and a distinction between 'primary' and 'secondary' artefacts cf. Wartofsky 1979, p. 201). With respect to workplace mathematics artefacts seem to play a twofold role: On the one hand, they are excellent places to hide mathematics by incorporating it into routines and algorithms which do not look like mathematics at all. Only when analysing the structure underlying the double entry bookkeeping or the algorithms implemented in sophisticated CAD software one can (re)discover the mathematics incorporated in these workplace practices. Consequently, studying workplace related mathematics for a large part may consist of analysing the artefacts used at the respective workplace. To me the papers by Bessot in section 3 and by Mercier in section 4 present prototypes of this approach - and they even offer terminological and factual progress in relying on work done in the sociology of work (cf. recent work by Rabardel (1995) for mathematical artefacts and instruments).

At present, the most important artefact seems to be modern information technology - and this very same technology seems to be the most effective way tc hide mathematics (cf. the banking study in Noss&Hoyles 1996). On the other hand, the same paper shows that this technology can be used to de-grey the black boxes, to show the mathematical relations and offer an opportunity to explore the inherent, implemented relations in a way, workplace reality would never allow because of the risk of material, financial and time losses (see already the argument at the end of part 2 of the conclusion). The practice of using sophisticated mathematics can be brought to the foreground and consciousness of the user by appropriate software and vocational training / education. It is modern computer technology and appropriate software which can be successfully used to really explore and understand the underlying (banking) mathematics.

More generally, modern information technology has an ambivalent role in the process of using mathematics at the workplace: It can be used as a way to hide mathematics in sophisticated software. Mathematics as a tool, a man-made artefact disappears in workplace routines - and modern technology can speed up this disappearance. On the other hand, the very same technology can be used in learning processes to foster understanding of the professional use of mathematics by explicitly modelling the hidden mathematical relations and offering software tools to explore and better understand the underlying mathematical models.

4. COMMENT ON RESEARCH METHODOLOGY

Looking into the four papers of the book grouped in section 4 on methodology, the most striking fact to me is that there is only work on mathematics at the workplace - and virtually nothing on teaching and learning processes of workplace related mathematics. Even if we leave the narrow confines of section 4 and have a closer look e.g. into section 3 on teaching future workers, we do not find a comment on research methodology for these teaching and learning processes. And the paper by Hogan&Morony in section 2 gives teachers a good start to find mathematics at the workplace - but does not comment on the teaching process where teachers should have special expertise. Consequently I start with a comment on the way to find and identify workplace related mathematics.

Section 4 presents two ways to do this: Zevenbergen vividly advocates the ethnographic case study at the workplace and Bessot show an example of this type of research. The paper by Mercier offers a somewhat different brand of the same kind of 'passive' approach: Studying the historical development of the use of mathematics (and mathematical instruments) within the workplace activities of fishermen and farmers raising cattle is an excellent prototype of an historical case study to identify mathematics used at work.

This set of papers in section 4 to me presents two insights worthwhile stating: Firstly, traditional survey and interview studies into the use of mathematics in workplace contexts tend to come up with no valuable information on the actual use of mathematics because mathematics is hidden from the perception of the interviewee by artefacts - be it material tools, workplace procedures or organisational features like distributing knowledge in a special way across a company's work hierarchy. In addition to that research into the use of mathematics in workplace contexts has to choose between participant observation in an ethnographic style / historical case studies (both more or less 'passive' methodologies; a lot more examples can be identified, especially in the Australian research community) or research has to create opportunities for the qualified worker to show her/his workplace practice including the mathematics therein (the more 'active' methodology; see for instance the examples in Noss&Hoyles (1996) or the study by Straesser&Bromme (1992) into technical drawing). The active approach nevertheless heavily depends on knowledge accumulated before in order to present the 'right' stimuli for the worker to come up with 'her/his' workplace mathematics. Both, active as well as passive research methods need a thorough and intensive analysis of data after the collecting of information to enable the researcher to (re)discover the mathematics hidden in the workplace practice. The superiority of ethnographic studies and stimulated response type of research over traditional interview and survey studies is deeply rooted in the characteristics of the use of mathematics at the workplace - namely the fact that the use of mathematics tends to be hidden from the perception of the worker.

To finalise these remarks on research methodology, I would like to reiterate and strengthen my introductory remark on research methodology: to date, there is no discussion on a special methodology for workplace related teaching and learning of mathematics. This could be an indication that the 'usual' research methodology for the analysis of teaching and learning processes can easily be adopted for workplace related mathematics. With 'legitimate peripheral participation' as at least one teaching and learning strategy for this type of mathematics, I doubt this easy way out of a necessary discussion which will come up soon.

5. REFERENCES

Chevallard, Y. (1991). La Transposition Didactique - du Savoir Savant au Savoir Enseignè, (2nd edition), Grenoble: La Pensée Sauvage.

Niss, M. (1994). Mathematics in Society. In R. Biehler, et al., (Eds.), Didactics of Mathematics as a Scientific Discipline (pp. 367-378). Dordrecht: Kluwer.

Noss, R., and Hoyles, C. (1996). The Visibility of Meanings: Modelling the Mathematics of Banking. International Journal of Computers for Mathematical Learning, 1(1), 3-31.

Nunes, T., Schliemann, A., and Carraher, D. (1993). Street Mathematics and School Mathematics. Cambridge: Cambridge University Press.

Rabardel, P. (1995). Les Hommes et la Technologie: Approche Cognitive des Instruments. Paris: Armand Colin.

Straesser, R., and Bromme, R. (1992). The Description of Solids in Technical Drawing - Results from Interviews of Experienced Draughtsmen. Paper presented at the Sixteenth PME Conference (vol. 3, p3-43 - 3-50), Durham, NH, USA.

Wartofsky, M. W. (1979). Models, Representation and Scientific Understanding. Dordrecht: Reidel.

INDEX OF SUBJECTS

INDEX OF AUTHORS

AUTHOR AFFILIATIONS

Annie Bessot
Didactique des Mathématiques
Laboratoire LEIBNIZ,
CNRS, Université Joseph Fourier, INPG
46, avenue Felix Viallet
38000 Grenoble
France
annie.bessot@imag.fr

Madeleine Eberhard
Didactique des Mathématiques
Laboratoire LEIBNIZ,
CNRS, Université Joseph Fourier, INPG
46, avenue Felix Viallet
38000 Grenoble
France
madeleine.eberhard@imag.fr

Jeff Evans
Mathematics & Statistics Group
Middlesex University Business School
Queensway
Enfield Middlesex EN3 4SF
United Kingdom
J.Evans@mdx.ac.uk

Susan L. Forman
Dept. of Mathematics and Computer Science
Bronx Community College
West 181 Street & University Ave.
Bronx, NY 10453-3102
United States of America
susan.forman@att.net

John Gillespie
School of Education,
University of Nottingham
Nottingham NG7 2RD
United Kingdom
John.Gillespie@nottingham.ac.uk

Corinne Hahn
NEGOCIA,
Chambre de Commerce et d'Industrie de Paris
8, avenue de la porte de Champerret
75838 Paris Cedex 17
France
chahn@schamp.ccip.fr

John Hogan
Centre for Curriculum and Professional Development
School of Education
Murdoch University
South Street
Murdoch WA 6150
Australia
John.Hogan@bigpond.com

Celia Hoyles
Institut of Education
University of London
20 Bedford way
London WC 1H 0AL
United Kingdom
choyles@ioe.ac.uk

Alain Mercier
INRP, Antenne de Marseille
UNIMECA, pôle technologique de Château-Gombert
60 rue Joliot Curie
13013 Marseille
France
mercier@univ-aix.fr

Will Morony
Australian Association of Mathematics Teachers Inc
GPO Box 1729
Adelaide SA 5001
Australia
wmorony@aamt.edu.au

Richard Noss
Institut of Education
University of London
20 Bedford way
London WC 1H 0AL
United Kingdom
rnoss@mentor.ioe.ac.uk

Stefano Pozzi
Department for Education and Employment
Caxton House
6-12 Tothill Street
London SW1H 9NA
United Kingdom
Stefano.POZZI@dfee.gov.uk

Jim Ridgway
School of Education
University of Durham
Leazes Road
Durham DH1 1TA
United Kingdom
Jim.Ridgway@durham.ac.uk

Lynn Arthur Steen
Department of Mathematics
St. Olaf College
1520 St. Olaf Avenue
Northfield, MN 55057-1098
United States of America
steen@stolaf.edu

Rudolf Straesser
IDM
University of Bielefeld
P.O. Box 100131
D-33501 Bielefeld
Germany
rudolf.straesser@uni-bielefeld.de

Gérard Vergnaud
Cognition et activités finalisées
CNRS, Université Paris VIII
2 rue de la Liberté
93 526 Saint-Denis Cedex 2
France
vergnaud@univ-paris8.fr

Geoff Wake
Centre for Mathematics Education
University of Manchester,
Oxford Street
Manchester M13 9PL
United Kingdom
Geoff_Wake@compuserve.com

Tine Wedege
Department of Mathematics and Physics
Roskilde University
Box 260
4000 Roskilde
Denmark
TIW@MMF.RUC.DK

Julian Williams
Centre for Mathematics Education
University of Manchester,
Oxford Street
Manchester M13 9PL
United Kingdom
JWILLIAMS@fs1.ed.man.ac.uk

Robyn Zevenbergen
Faculty of Education
Griffith University
PMB 50, GCMC
Bundall, QLD 9726
Gold Coast, Australia
R.Zevenbergen@eda.gu.edu.au

Pieter van der Zwaart
SLO, Institute for Curriculum Development
P.O. Box 2041
7500CA Enschede
The Netherlands
P.vanderZwaart@slo.nl

Mathematics Education Library

Managing Editor: A.J. Bishop, Melbourne, Australia

1. H. Freudenthal: *Didactical Phenomenology of Mathematical Structures.* 1983
ISBN 90-277-1535-1; Pb 90-277-2261-7
2. B. Christiansen, A. G. Howson and M. Otte (eds.): *Perspectives on Mathematics Education.* Papers submitted by Members of the Bacomet Group. 1986.
ISBN 90-277-1929-2; Pb 90-277-2118-1
3. A. Treffers: *Three Dimensions.* A Model of Goal and Theory Description in Mathematics Instruction The Wiskobas Project. 1987 ISBN 90-277-2165-3
4. S. Mellin-Olsen: *The Politics of Mathematics Education.* 1987
ISBN 90-277-2350-8
5. E. Fischbein: *Intuition in Science and Mathematics.* An Educational Approach. 1987
ISBN 90-277-2506-3
6. A.J. Bishop: *Mathematical Enculturation.* A Cultural Perspective on Mathematics Education. 1988 ISBN 90-277-2646-9; Pb (1991) 0-7923-1270-8
7. E. von Glasersfeld (ed.): *Radical Constructivism in Mathematics Education.* 1991
ISBN 0-7923-1257-0
8. L. Streefland: *Fractions in Realistic Mathematics Education.* A Paradigm of Developmental Research. 1991 ISBN 0-7923-1282-1
9. H. Freudenthal: *Revisiting Mathematics Education.* China Lectures. 1991
ISBN 0-7923-1299-6
10. A.J. Bishop, S. Mellin-Olsen and J. van Dormolen (eds.): *Mathematical Knowledge: Its Growth Through Teaching.* 1991 ISBN 0-7923-1344-5
11. D. Tall (ed.): *Advanced Mathematical Thinking.* 1991 ISBN 0-7923-1456-5
12. R. Kapadia and M. Borovcnik (eds.): *Chance Encounters: Probability in Education.* 1991 ISBN 0-7923-1474-3
13. R. Biehler, R.W. Scholz, R. Sträßer and B. Winkelmann (eds.): *Didactics of Mathematics as a Scientific Discipline.* 1994 ISBN 0-7923-2613-X
14. S. Lerman (ed.): *Cultural Perspectives on the Mathematics Classroom.* 1994
ISBN 0-7923-2931-7
15. O. Skovsmose: *Towards a Philosophy of Critical Mathematics Education.* 1994
ISBN 0-7923-2932-5
16. H. Mansfield, N.A. Pateman and N. Bednarz (eds.): *Mathematics for Tomorrow's Young Children.* International Perspectives on Curriculum. 1996
ISBN 0-7923-3998-3
17. R. Noss and C. Hoyles: *Windows on Mathematical Meanings.* Learning Cultures and Computers. 1996 ISBN 0-7923-4073-6; Pb 0-7923-4074-4

18. N. Bednarz, C. Kieran and L. Lee (eds.): *Approaches to Algebra*. Perspectives for Research and Teaching. 1996 ISBN 0-7923-4145-7; Pb 0-7923-4168-6
19. G. Brousseau: *Theory of Didactical Situations in Mathematics*. Didactique des Mathématiques 19701990. Edited and translated by N. Balacheff, M. Cooper, R. Sutherland and V. Warfield. 1997 ISBN 0-7923-4526-6
20. T. Brown: *Mathematics Education and Language*. Interpreting Hermeneutics and Post-Structuralism. 1997 ISBN 0-7923-4554-1
21. D. Coben, J. O'Donoghue and G.E. FitzSimons (eds.): *Perspectives on Adults Learning Mathematics*. Research and Practice. 2000 ISBN 0-7923-6415-5
22. R. Sutherland, T. Rojano, A. Bell and R. Lins (eds.): *Perspectives on School Algebra*. 2000 ISBN 0-7923-6462-7
23. J.-L. Dorier (ed.): *On the Teaching of Linear Algebra*. 2000 ISBN 0-7923-6539-9
24. A. Bessot and J. Ridgway (eds.): *Education for Mathematics in the Workplace*. 2000 ISBN 0-7923-6663-8

KLUWER ACADEMIC PUBLISHERS – DORDRECHT / BOSTON / LONDON

The manufacturer's authorised representative in the EU is Springer Nature Customer Service Centre GmbH, Europaplatz 3, 69115 Heidelberg, Germany. If you have any concerns regarding our products, please contact ProductSafety@springernature.com

Printed and bound by CPI Group (UK) Ltd, Croydon, CR0 4YY
30/09/2026
02225494-0001